Second Grade Math

For Home School or Extra Practice

By Greg Sherman

Home School Brew Press

www.HomeSchoolBrew.com

© 2013. All Rights Reserved.

Cover Image © nikoniko_happy - Fotolia.com

Table of Contents

Number Sense 5

Addition/Subtraction 14

Fractions 24

Operations 36

Money 46

Shapes/Patterns 53

Length 62

Weight/Capacity 70

Probability 80

Graphs 90

Answer Key 103

Disclaimer

This book was developed for parents and students of no particular state; while it is based on common core standards, it is always best to check with your state board to see what will be included on testing.

About Us

Homeschool Brew was started for one simple reason: to make affordable Homeschooling books! When we began looking into homeschooling our own children, we were astonished at the cost of curriculum. Nobody ever said homeschool was easy, but we didn't know that the cost to get materials would leave us broke.

We began partnering with educators and parents to start producing the same kind of quality content that you expect in expensive books...but at a price anyone can afford.

We are still in our infancy stages, but we will be adding more books every month. We value your feedback, so if you have any comments about what you like or how we can do better, then please let us know!

To add your name to our mailing list, go here: http://www.homeschoolbrew.com/mailing-list.html

Number Sense

One of the first skills we develop as children is counting. Everyone enjoys asking toddlers how old they are and how many fingers they can count. We are constantly counting things, adding and subtracting items, multiplying and dividing quantities during our average day. During third grade, students will learn several new skills with numbers. Some of these skills include how to name numbers, where to position numbers to perform various operations, and specific traits of special numbers, like even and odd numbers.

Numbers from one and higher are classified as Whole Numbers. Zero is an important number, but functions mainly as a placeholder. Numbers between zero and one are composed of fractions and because they have values less than one, they are not defined as Whole Numbers.

To be able to work with Whole Numbers, the students need to learn how to name them based on how many "places" are occupied. For example, the number ten (10) is named as such because it has numbers occupying the ones and the tens places. The table below shows the place values and basic names for Whole Numbers up to One Hundred Thousand:

Ones	Tens	Hundreds	Thousands
1	10	100	1,000

Naming Numbers

Writing Names are given according to the number of "places" the number occupies. For example, 10 is named "ten" because the biggest place it occupies is the tens place.

- One Hundred (100) is named because it occupies up to the hundreds place.
- Two Hundred is named for the number two (2) and the number of places the whole number occupies (200).
- Five Thousand is named for the number five (5) and the number of places the whole number occupies (5,000).

To make naming numbers more interesting, the following examples include numbers in the various places instead of zeros:

- 2,124   Name: Two thousand, one hundred twenty-four
- 357   Name: Three hundred fifty-seven
- 17   Name: Seventeen

When zeros are used as place markers, they are NOT part of the name:

- 102   Name: One hundred two
- 3,052   Name: Three thousand, fifty-five

The above numbers are called Writing Numbers. They are how whole numbers are written when using them in an equation or problem or as a basic description of the number.

Ordinal Names

Another naming process for numbers is using the numbers themselves as place markers. These number names are used for ranking items, ordering number placements in a list, or indicating where numbers are located on a number line relative to each other.

⟵ 1 2 3 4 5 6 7 8 9 10 11 12 13 14 15 16 17 18 19 20 21 22 23 24 25 ⟶

The number line above shows whole numbers from One to Twenty-Five. Ordinal naming of the numbers on this number line looks like the following:

FIRST	SECOND	THIRD	FOURTH	FIFTH	SIXTH	SEVENTH	EIGHTH	NINTH	TENTH
1st	2nd	3rd	4th	5th	6th	7th	8th	9th	10h

Using the number line and Ordinal placement names, the following tasks can be performed:

1. What number is the 3rd to the right of 20?

2. If Lisa came in 3rd in the running race, what were the places that came in before?

3. Which number place comes after 36th place?

4. Which place is five places to the left of 14?

Even and Odd Numbers

Even numbers are all multiples of the number two. Or, they are all divisible by two. If a number cannot be divided completely by two (into two equal halves), it is NOT an even number.

Counting by even numbers always results in the next number in line also being even. The following examples are counting using even numbers:

Two's: 2, 4, 6, 8, 10, 12, 14, 16, 18, 20
Four's: 4, 8, 12, 16, 20, 24, 28, 32, 36
Sixes: 6, 12, 18, 24, 30, 36, 42, 48, 54

If a number cannot be completely dived into two equal parts, it is an odd number. In other words, if it is not even, it is odd.

When counting by an odd number, like three, the numbers in the order are presented in a pattern of odd, even, odd, even, etc. The following examples are counting using odd numbers:

Three's: 3, 6, 9, 12, 15, 18, 21, 24, 27, 30

Five's: 5, 10, 15, 20, 25, 30, 35, 40, 45, 50

Seven's: 7, 14, 21, 28, 35, 42, 49, 56, 63, 70

Comparing Numbers

When comparing numbers, students may use a number line or chart to help them understand where each number is in relation to the others. For example, is the number in question greater than or less than another number in the chart or line? When counting by numbers, which number comes next in the order? The best way to visualize how numbers relate to one another is to study a number chart. The one below is up to One Hundred:

1	2	3	4	5	6	7	8	9	10
11	12	13	14	15	16	17	18	19	20
21	22	23	24	25	26	27	28	29	30
31	32	33	34	35	36	37	38	39	40
41	42	43	44	45	46	47	48	49	50
51	52	53	54	55	56	57	58	59	60
61	62	63	64	65	66	67	68	69	70
71	72	73	74	75	76	77	78	79	80
81	82	83	84	85	86	87	88	89	90
91	92	93	94	95	96	97	98	99	100

By using the above chart is becomes easy to answer a few questions:

1. Which number is 5 less than 23?

    By moving 5 numbers back from 23, you get to 18.

2. How many numbers greater is 100 from 84?

    By counting how many numbers are between 84 and 100, you get an answer of 16.

3. If you move 37 places past 22, what number do you land on?

By counting 37 places past 22, you land on 59.

4. Which number is between 65 and 67?

    By looking at the chart, the student finds 66 is between 65 and 67.

Rounding Numbers

Sometimes to make certain number operations easier for the student, they are first required to "round" the numbers to the simplest amount with regard to place values. The basic rule for rounding up or down is this: if the number is 4 or less you round down, if the number is 5 or greater you round up. The following numbers are examples of "rounding to the nearest tens place value:

Number	Rounded Number to Tens Place	Explanation
14	Round down to 10	14 is closer to 10 than 20
27	Round up to 30	27 is closer to 30 than 20
42	Round down to 40	42 is closer to 40 than 50
57	Round up to 60	57 is closer to 60 than 50

Number Sense – Practice Sheet

1. What is the place value for the number three in 3,526?

2. What is the place value for the number 7 in 743?

3. What is the place value for the zero in 107?

4. Circle the whole numbers: 3,   4.3,   6,   ½,   7.1,   9/10

5. True or False: 4 3/4 is a whole number.

Number Names

1. Write the name for 205.

2. Write the name for 1,706.

3. Write the number for Three thousand, fifty-two.

4. Write the name for 47th

5. Write the number for Eighty-Eighth

6. Round 47 to the closest Tens Place

Odds and Evens

1. Is 237 odd or even?

2. Is 425 odd or even?

3. Is 562 odd or even?

4. If you count by three's are all the numbers odd?

5. If you count by four's are all the numbers even?

Comparing Numbers

1. Is 51 greater than or less than 15?

2. Which is the 3rd number after 10?

3. Which five numbers are between 12 and 18?

4. How many zeros are in One Thousand?

5. How many tens (count by tens) are between Twenty and Fifty?

Number Sense - Quiz

1. Write the name for 2,358

2. Write the name for 637

3. Round 84 to the nearest Tens Place:

4. Write the number for Seventy-Ninth

5. Which numbers are in between 23 and 27?

6. Which number is even: 14 or 17

7. Which number is greater: 23 or 32

8. Fill in the missing number: 2   4   _____   8   10

9. If you come in 7th place in a race, how many people came in to the finish before you?

10. Which number is less than 27: 26 or 28

Addition/Subtraction

During Second Grade, students learn how to add and subtract in a specific order:

- adding and subtracting single-digit numbers
- adding and subtracting double digit numbers
- adding and subtracting triple-digit numbers

Adding and Subtracting Single-Digit Numbers

A quick review of adding single digit numbers:

1 + 1 = 2	2 + 2 = 4	3 + 3 = 6	4 + 4 = 8	5 + 5 = 10	6 + 6 = 12
1 + 2 = 3	2 + 3 = 5	3 + 4 = 7	4 + 5 = 9	5 + 6 = 11	6 + 7 = 13
1 + 3 = 4	2 + 4 = 6	3 + 5 = 8	4 + 6 = 10	5 + 7 = 12	6 + 8 = 14
1 + 4 = 5	2 + 5 = 7	3 + 6 = 9	4 + 7 = 11	5 + 8 = 13	6 + 9 = 15

A quick review of subtracting single-digit numbers:

2 − 1 = 1	4 − 2 = 2	6 − 3 = 3	8 − 4 = 4	10 − 5 = 5	12 − 6 = 6
3 − 2 = 1	5 − 3 = 2	7 − 4 = 3	9 − 5 = 4	11 − 6 = 5	13 − 7 = 6
4 − 3 = 1	6 − 4 = 2	8 − 5 = 3	10 − 6 = 4	12 − 7 = 5	14 − 8 = 6
5 − 4 = 1	7 − 5 = 2	9 − 6 = 3	11 − 7 = 4	13 − 8 = 5	15 − 9 = 6

From adding and subtracting single-digit numbers came the concept and practice of learning *fact families* to help students add and subtract more quickly.

Some Fact Families have been started in the above review of adding and subtracting single-digit numbers:

Example 1

1 + 2 = 3   2 + 1 = 3   3 – 1 = 2   3 – 2 = 1

Example 2

1 + 3 = 4   3 + 1 = 4   4 – 1 = 3   4 – 3 = 1

Example 3

1 + 4 = 5   4 + 1 = 5   5 – 1 = 4   5 – 4 = 1

By memorizing the process of adding two numbers that give the same sum and then when each number is subtracted from that sum it results in the original numbers again, gives the students a pattern they can use to take tests and compute single-digit operations faster.

Basically, it is easier to memorize the relationship between three numbers than trying to memorize each and every individual operation between random numbers.

Adding and Subtracting Double-Digit Numbers

When adding with double-digit numbers, Second Grade students will learn a process called "regrouping." This process was previously taught as "carrying the one" or "carrying the extra or remainder." The other concept the students will learn is to line up the proper place columns. If the numbers being added are not property aligned, the wrong numbers will be added and the student will arrive at an incorrect solution.

Students first begin adding numbers based in ten: 10, 20, 30, 40, etc.

10	20	40	10	30	60	20
+ 20	+ 30	+ 20	+ 10	+ 30	+ 50	+ 50
30	50	60	20	60	110	70

Next, students add double-digit and single-digit numbers without Regrouping:

12	18	23	33	76	47	52
+ 3	+ 1	+ 4	+ 5	+ 2	+ 1	+ 7
15	19	27	38	78	48	59

Adding using Regrouping is then introduced and practiced by the Second Grade students. Regrouping is the process used when the first numbers added produce a double-digit sum. The number in the ones place is written in the sum and the number in the tens place is "regrouped" with the number in the tens place being added.

Example:

```
(regrouped)  1
             25
           +  7
           ----
             32
```

Since 7 + 5 = 12, the 2 is placed in the ones place in the sum and the 1 is Regrouped with the 2 in the tens place for 25 giving the correct answer of 32.

The following double-digit addition examples all use Regrouping:

```
  1        1        1        1        1        1        1
 27       32       45       85       76       15       57
+ 3      + 9      + 6      + 5      + 8      + 9      + 7
───      ───      ───      ───      ───      ───      ───
 30       41       51       90       84       24       64


  1        1        1        1        1        1        1
 12       28       23       33       76       47       52
+19      +13      +28      +39      +36      +27      +29
───      ───      ───      ───      ───      ───      ───
 31       41       51       72      112       74       81
```

Subtracting Double-Digit numbers sometimes uses a process called "Borrowing." It is similar in concept to Regrouping, but instead of adding the one to the Regrouped tens place, a value of "10" is moved to the ones place from the tens place of the larger number to allow the subtraction operation to work.

For Example:

```
  25              20 + (5)            10 + (15)
-  7      →       -   7       →       -    7
 ───              ────────            ─────────
  18                  18                  18
```

The above example shows that since the 5 is too small to subtract 7, the 5 "Borrows" a "10" from the "20" to make 15. The 20 is reduced to a 10 and now the 7 can be subtracted from 15 which gives the answer of 8. The 10 is brought down into the answer to give the correct answer of 18.

As with addition, Borrowing is not always necessary. The following examples show subtraction problems that both need to use Borrowing and those that don't:

```
  27       32       45       85       76       15       57
-  3     -  9     -  6     -  5     -  8     -  9     -  7
  24       23       39       80       68        6       50

  57       28       83       43       76       47       52
- 19     - 13     - 28     - 39     - 36     - 25     - 39
  38       15       55        4       40       23       13
```

Adding and Subtracting Triple-Digit Numbers

Adding and subtracting triple-digit numbers is essentially the same as adding and subtracting double-digit numbers with respect to Regrouping and Borrowing, but it sometimes needs an extra step to complete the process.

Borrowing, for example, requires an extra step:

```
  251             200 + 50 + 1           100 + 140 + 11
-  67           -       67             -        60 +  7
  184                   184                          184
```

The above example shows that since the 1 is too small to subtract 7, the 1 "Borrows" a "10" from the "50" to make 11. The 50 is reduced to a 40 and now the 7 can be subtracted from 11 which gives the answer of 4. Now, 40 is too small to subtract 60, so the 40 Borrows 100 from 200 to make 140. When 60 is subtracted from 140 the answer is 80. The 100 is brought down into the answer to give the correct answer of 184.

Otherwise, the following triple-digit number problems are just like single-digit and double-digit number problems:

327	532	745	285	177	619	757	984
+ 4	+ 7	+ 3	+ 5	- 8	- 9	- 4	- 7
331	539	748	290	169	610	753	977

157	328	183	343	576	647	852	943
+ 19	+ 13	+ 28	+ 39	- 15	- 25	- 41	- 67
176	341	211	382	561	623	711	876

225	208	873	543	766	497	952	457
+ 129	+ 413	+ 628	+ 137	- 236	- 325	- 839	- 125
354	621	1,501	680	530	172	113	332

When subtracting numbers with zeros, the borrowing begins with the first zero rather than the end of the number as with non-zero numbers.

So, when borrowing to complete this subtraction problem, it starts with the first zero borrowing 100 and then the second zero borrowing 10. Then the problem looks like this:

$$\begin{array}{r} 200 \\ -125 \\ \hline 75 \end{array} \quad \longrightarrow \quad \begin{array}{r} 100+90+10 \\ -\ 100+20\ +5 \\ \hline 0+70\ +5 \end{array}$$

$$\begin{array}{r} 205 \\ -129 \\ \hline 76 \end{array} \quad \begin{array}{r} 200 \\ -113 \\ \hline 87 \end{array} \quad \begin{array}{r} 870 \\ -628 \\ \hline 242 \end{array} \quad \begin{array}{r} 503 \\ -137 \\ \hline 366 \end{array} \quad \begin{array}{r} 706 \\ -236 \\ \hline 470 \end{array} \quad \begin{array}{r} 490 \\ -325 \\ \hline 165 \end{array} \quad \begin{array}{r} 900 \\ -839 \\ \hline 61 \end{array} \quad \begin{array}{r} 450 \\ -125 \\ \hline 325 \end{array}$$

Addition and Subtraction - Practice Sheet

Fill in the missing numbers:

1. 1 + 3 = 4

    3 + ___ = 4

    4 − 1 = 3

    4 − ___ = 1

2. 2 + 5 = 7

    ___ + 2 = 7

    7 − ___ = 2

    7 − ___ = 5

Solve (remember to line up proper places or columns to get the correct answer)

1.	12 + 1 =	2.	32 + 84 =	3.	810 + 21 =
4.	90 − 12 =	5.	98 − 15 =	6.	992 − 827 =
7.	85 + 95 =	8.	410 + 127 =	9.	321 + 620 =
10.	87 − 18 =	11.	726 − 552 =	12.	565 − 221 =
13.	97 + 12 =	14.	123 + 456 =	15.	722 + 32 =

Greater Than, Less Than, Equal To

1.   12 + 14 + 1 _____ 30 – 3

2.   125 – 18 _____ 100 + 2 + 3

3.   715 – 113 _____ 300 + 200 + 157

Addition and Subtraction - Quiz

Fill In the Blanks

1. 12 + _____ = 17, _____ + 12 = 17, 17 - _____ = 12, 17 – 12 = _____

2. 23 + _____ = 50, _____ + 23 = 50, 50 - _____ = 23, 50 – 23 = _____

Solve

1. 871 + 123 =

2. 726 – 13 =

3. 35 + 27 =

4. 872 − 773 =

5. 73 + 271 =

6. 956 − 897 =

7. 25 + 29 =

8. 621 − 599 =

Fractions

Learning fractions and their functions opens a new world of math and its various uses for the Second Grade student. After learning some of the properties of fractions, students will be able to identify their uses in our daily lives.

When we bake or cook, the measurements are generally in fractions. For example, when baking cookies, the recipe usually calls for one-half cup of sugar, three-quarters cup of brown sugar, one-half cup butter, etc. Many of the tools used to fix cars and other items around the house are calculated in fractions. It is common to find a 3/4 inch wrench or a 5/16 socket in a basic home toolbox.

Fractions are numbers that are less than one, but greater than zero. These numbers all have values between zero and one. Some are very close in value to one like the fraction 9/10. They are equal parts of the Whole Number One.

Fractions are composed of two different numbers. The top number is called the numerator and the bottom number is called the denominator. The numerator represents part of the total and the denominator is the total amount represented.

$$\frac{\text{NUMERATOR}}{\text{DENOMINATOR}}$$

It is easier to see what a fraction really represents by using lists of items and pie charts. These visual tools show what the whole amount looks like and how the fraction covers only a part of the whole.

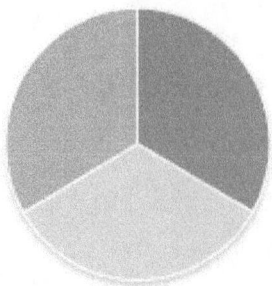

The above picture shows three equal sections of a pie. Since the total number of sections is three, the denominator of the fraction for this picture is 3. The numerator can be 1, 2, or 3 depending on what question is being asked:

1. What is the fraction of the circle colored the lightest gray? 1 out of 3 so, 1/3 is the fraction.

2. What is the fraction for the lightest section and the darkest sections together? 2 out of 3 sections are described, so 2/3 is the fraction.

3. What fraction represents the entire circle? Since there are three sections, and the entire pie is composed of three sections, the fraction is 3/3.

Other examples of visual fraction pictures are presented below:

■ ■ ■ ■ ☐   The total number of squares is 5, so the denominator is 5.

What fraction of 5 is the white square? Since there is only one white square, the fraction is 1 out of 5 or 1/5. AND, if the one white square is 1/5, the 4 remaining black squares are the fraction 4/5.

▲▲▲▲△△   The total number of triangles is 6, so the denominator is 6.

What are the fractions that represent the number of white triangles out of 6 and the number of black triangles out of 6? There are 2 white triangles, so their fraction is 2/6 and there are 4 black triangles, so their fraction is 4/6.

Equal Halves

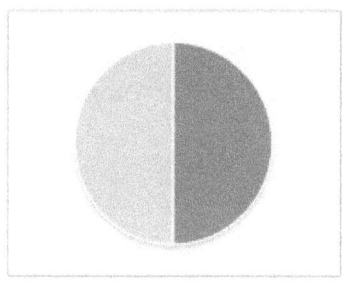
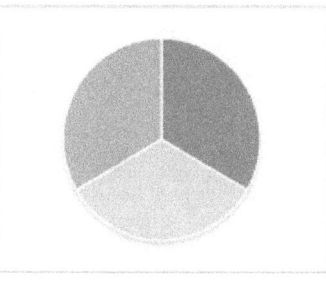
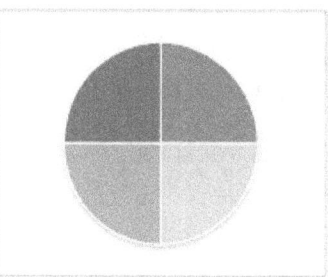

Equal Halves   Equal Thirds   Equal Fourths
(2 parts of 2)   (3 parts of 3)   (4 parts of 4)

The above pies show equal parts of the whole. Each pie is the same size, but can be divided in many different ways and create many different kinds of fractions. One piece of each pie is represented as 1 part of the whole number of pieces. So, one part of two is 1/2, one part of three is 1/3, and one part of four is 1/4.

IT IS VERY IMPORTANT TO REMEMBER that the larger the denominator, the smaller the fraction. For example, as demonstrated by the above pies, 1/4 is smaller than 1/3, which is smaller than 1/2.

Based on the above information, the following comparisons can be made:

1. Greater Than or Less Than: 5/6 > 5/8 > 5/10 AND 3/10 < 3/8 < 3/4   because the larger the denominator, the smaller the individual pieces of the whole.

2. Equivalent Fractions: Using the pie charts above, which two fractions are equal?   If the charts are compared based on the area covered by the   individual pieces of   each chart, it can be

seen that two of the Equal Fourths cover the same area as one of the Equal Halves. Therefore, 2/4 and 1/2 are equivalent fractions.

3. Fractions that Equal One:

If the denominator is the total number of parts in the whole and the numerator is a portion of the whole number, then if the numerator and denominator are equal, all the parts of the total are represented and the fraction equals the whole.

So, 1 = 2/2 = 3/3 = 4/4 = 5/5 … 100/100 etc.

The following problems are examples of what Second Grade students will be learning:

1. Color in 2/5 of the circle below:

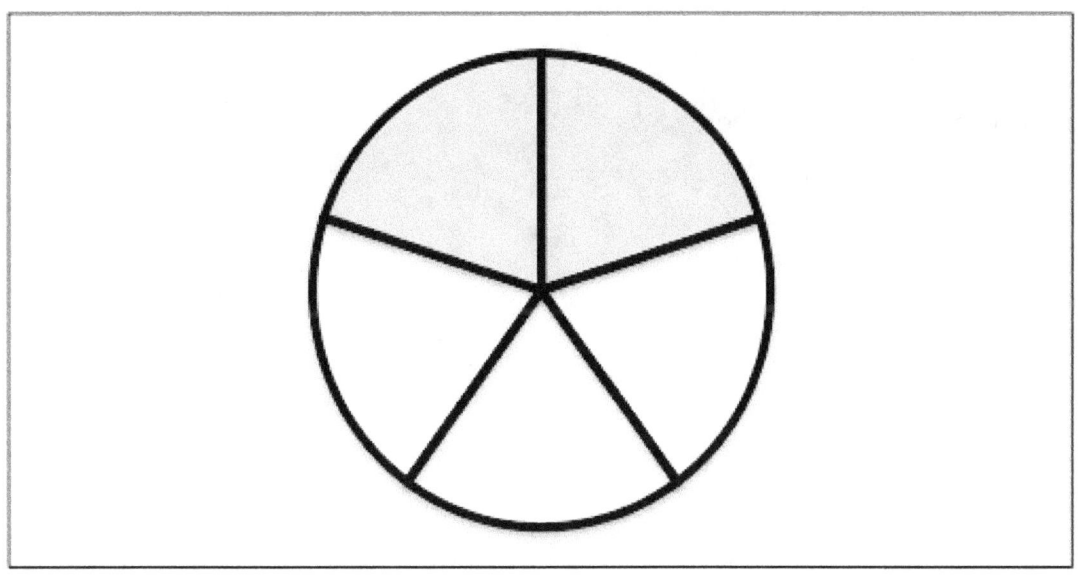

2. Color in half of the sections (3/6)

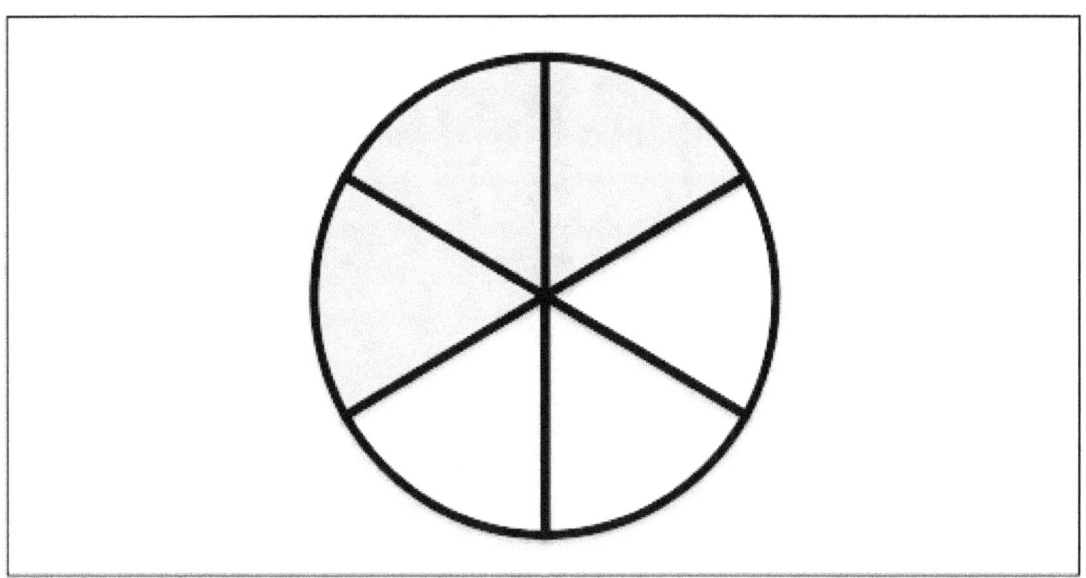

3.

How many lightening bolts are there?   7

How many are shaded?   5

How many are white?   2

What is the fraction of lightening bolts that are shaded?   5/7

What is the fraction of lightening bolts that are white? 2/7

What is the fraction that represents all of the lightening bolts?   7/7 = 1

4.

How many total shapes?   5

How many suns?   3

How many moons?   2

What fraction of the total shapes are the suns?   3/5

What fraction of the shapes are moons?   2/5

5. Based on the graphs in Questions 1 and 2, which is bigger: 2/5 or 3/6?

   Answer: 3/6

6. Indicate in the following sets of fractions, which are greater than, lesser than, or   equal:

   3/3 __=__ 7/7   3/5 _____>_____ 3/8   99/100 _____>___ 98/100

   23/25 ____<___24/25   1/2 _____<____2/2   4/4 _____=____8/8

7. Indicate what fraction the shaded objects represent:

       = 3/4   = 5/6

8.

    Fill in the blanks

         Fill in the   blanks

    1/2 1/3 1/4 [   ]1/6 1/7 Answer: 1/5

    1/8 2/8 3/8 4/8 [   ] 6/8 7/8 8/8 Answer: 5/8

    1/1 2/2 [   ] 4/4 5/5 6/6 7/7 Answer: 3/3

9. Which row from number 8 has all the fractions equal to 1?   Answer: Row 3

Elementary students will practice and learn fractions throughout their educational experience. Practice solving problems like those presented above, will help students approach learning fractions with more confidence and assurance, which makes fractions appear less complicated and intimidating.

Fractions – Practice Sheet

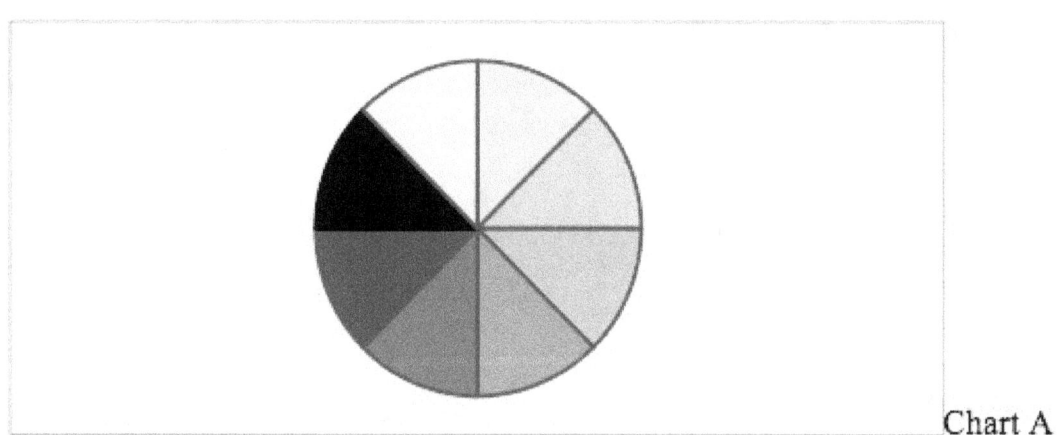

Chart A

1. How many pieces total are there in the above pie chart?

2. If you put the white piece and the darkest piece together what would be the fraction?

3. How many pieces equal 1/2 of the pie?

4. What is the fraction for just one piece?

5. If you picture the pie chart as a clock face, how many pieces equal 1/4 hour (15 min)?

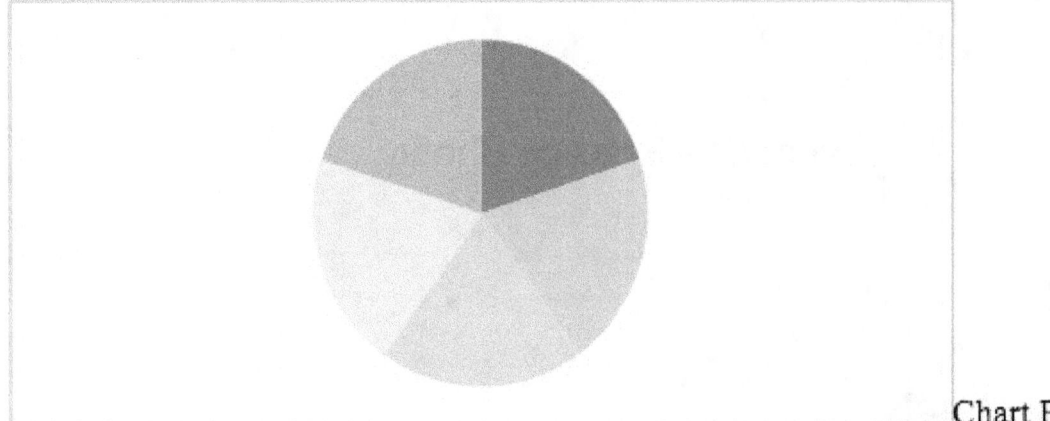
Chart B

6. Is one piece from Chart A bigger or smaller than one piece from Chart B?   [smaller]

7. Which is bigger, 1/8 or 1/5?

Answer the following questions using Chart A and Chart B:

8. Which is bigger, 3/5 or 4/8?

9. If each chart were a real pie, which one would feed more guests?

10. What is the denominator for Chart A?   Chart B?

Answer the following questions using the above set of shapes:

36

1. What is the total number of shapes?

2. How many squares are there?

3. Write the number of squares and the total number of shapes as a fraction:

4. How many shapes are not circles?

5. What fraction of the total shapes are circles?

6. What fraction of the total shapes are triangles?

Compare Fractions

7. Put these fractions in order of smallest to largest: 2/3 2/7 2/5

8. Indicate whether the fractions are Greater Than, Less Than, or Equal: 7/7 ___ 3/3

9. Which Numerator is Greater: 5/16 or 9/10

10. Which Denominator is Greater: 1/8 or 4/7

Fractions - Quiz

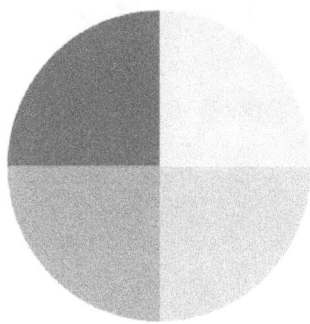

1. How many pieces are there in the above pie chart?

2. What fraction is the lightest color?

3. How many pieces make 1/2 of the pie chart?

4. What is the fraction for 3 of the four pieces?

5. What denominator should be used for the above set of lightening bolts?

6. What is the fraction for the white lightening bolt?

7. What fraction of the lightening bolts are gray?

8. What fraction of the lightening bolts are black?

Compare

9. Which is greater: 7/8 or 1/9

10. How many of the fraction 1/16 are needed to make a fraction equal to 1?

Operations

Other than the traditional math operations, like addition and subtraction, other types of math operations will be taught in Second Grade. These operations include, understanding place values, estimating and rounding numbers based on place values, learning time and calendar functions, and the beginnings of multiplication and division.

Number Places

Second Grade students will learn number places up to One Thousand. The number places are as follows:

1s 1, 2, 3, 4, 5, 6, 7, 8, and 9

10s 10, 20, 30, 40, 50, 60, 70, 80, and 90

100s 100, 200, 300, 400, 500, 600, 700, 800, and 900

1000s 1,000   2,000   3,000   4,000   5,000   6,000   7,000   8,000   and 9,000

The various zeros in the 10s, 100s, and 1000s places are replaced with standard numbers to create the whole numbers students will use in their math operations. The following examples show how students will practice and learn place values:

1. What is the place value for the number 7?

1,007 ones

1,271 tens

1,788 hundreds

7,102 thousands

2. Write out the following numbers into their expanded forms

25 = 20 + 5

146 = 100 + 40 + 6

2,789 = 2,000 + 700 + 80 + 9

3. Regroup the following expanded numbers

4 tens + 18 ones = 5 tens + 8 ones

3 tens + 25 ones = 5 tens + 5 ones

3 hundreds + 13 tens + 6 ones = 4 hundreds + 3 tens + 6 ones

8 hundreds + 24 hundreds + 12 ones = 10 hundreds + 5 hundreds + 2 ones

4. Convert from expanded numbers to regular numbers

100 + 30 + 5 =   135

20 + 7 =   27

3,000 + 400 + 50 + 9 =   3,459

Time and Calendar Functions

The above clocks are called analog clocks because they do not use digital technology to display the time. Digital clocks display the time exactly as it is written by using numbers to represent the hours and minutes separated by a colon.

For example, the above times written in standard time format (also digital) look like this:

      2:00   2:15   2:30   2:45

One piece of information that analog clocks do not display is whether the time is PM (Noon to Midnight) or AM (Midnight to Noon). So, when reading the time from an analog clock, the student will need to have AM or PM information provided in the problem.

Times that occur on the hour, the big hand (minute hand) on the 12 and the little hand (hour hand) on the number can be written with an O'clock descriptor. So, the above 2:00 can also be written as 2 O'clock. PM and AM still need to be added to determine day or night distinctions.

The above times are also learned as fractions of time:

2:15 (AM or PM) =   Quarter after Two, Quarter past Two, and 1/4 past the hour

2:30 (AM or PM) =   Half (1/2) past Two, Half (1/2) past the Hour

2:45 (AM or PM) =   Quarter to Three and 1/4 to Three

Students also learn that there are 24 hours in a day (AM and PM combined) and that although analog clocks are set for standard time, digital clocks and written time can be expressed in military time or by using a 24-hour clock. They will only need to learn Standard Time.

Learning to read and use a calendar are also skills students are taught in Second Grade. Math elements are involved because the standard calendar is divided into days, weeks, and months and each has a numerical value. Students learn that there are seven days in one week and generally thirty days per month. They learn more specifically that certain months have 30 days, some have 31 days, and that February is tricky because is although it usually has 28 days, every four years has 29 days.

If the students are told that the week begins on Sunday and ends on Saturday they can answer some basic calendar questions:

1. Which day of the week is three from Monday?   Thursday

2. How many days are not on the Weekend?   Five

3. How many days are in a week?   Seven

4. Which day is considered to be the middle day?   Wednesday

5. Which day of the week begins with the letter F?   Friday

6 How many days begin with the letter T?   Two

7. What days of the week do you go to school?   Monday, Tuesday, Wednesday, Thursday, and Friday

Students also learn the order of the 12 months on the calendar:

January	**2013/2014**						
February							
March	S	M	T	W	T	F	S
April							
May	1	2	3	4	5	6	7
June							
July	8	9	10	11	12	13	14
August	15	16	17	18	19	20	21
September							
October	22	23	24	25	26	27	28
November							
December	29	30	(31)				

Students will be expected to know the following facts about the months of the year:

- January 1 is the first day of each year and is called New Year's Day
- During a Leap Year, February has 29 days instead of 28.
- The year is divided into 4 seasons: Winter, Spring, Summer and Fall.
- The end of each year is December 31 and is called New Year's Eve.

Multiplication and Division

Students will learn multiplication tables up to 10 and be able to divide numbers up to 10. They will first learn how to count groups of numbers:

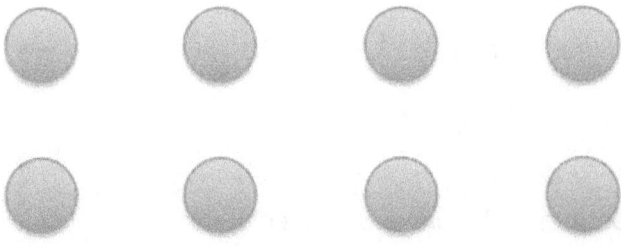

1. How many circles total?   8

2. How many rows of 4 are there?   2

3. If you add 2 rows of 4, how many circles do you have?   8

4. How many columns of 2 are there? 4

5. If you add 4 rows of 2, how many circles do you have?   8

So, if you add 2 together 4 times or you add 4 together 2 times each answer is 8. Therefore, 2 times 4 and 4 times 2 both equal the total number of 8 circles.

Once the Second Grade students have a basic understanding of how multiplication is related to addition, they can begin memorizing their multiplication tables.

Using the above circles again, students can begin learning division:

1. How many circles total?   8

2. How many rows of 4 are there?   2

3. How many columns of 2 are there? 4

So, the total number of 8 circles can be divided into 2 groups of 4 or 4 groups of 2.

Operations – Practice Sheet

Place Values

1. What is the place value of the 8 in 185?

2. What is the place value of the 9 in 19?

3. What is the place value of the 2 in 207?

4. What is the place value of the 4 in 4,123?

Expanded Numbers

1. Write out the expanded number for 352:

2. Write out the expanded number for 85:

3. Write out the expanded number for 1,278:

4. Write the standard number for 20 + 9:

5. Write the standard number for 300 + 70 + 3:

6. Write the standard number for 5,000 + 200 + 20 + 1:

The clock above shows the hour hand on the 2 and the minute hand on the 12. The time represented is 2:00.

1. If it is during the day, is the time 2:00 AM or 2:00 PM?

2. If the minute hand is moved to the 6, is the time 2:15 or 2:30?

3. If the hour hand is moved to the 6, and it is after midnight but before noon, what time is it?

4. If the time is 8:15 PM, is it also 1/4 after 8:00 PM or 1/4 to 8:00 PM?

Calendar

1. Which day of the week begins with the letter W?

2. How many days are in the weekend?

3. Which days are the weekend days?

4. Which day is next after Thursday?

Multiplication and Division

1. If there are 3 groups of 5 apples to be shared, how many people get an apple?

2. If there are 25 pencils and 5 students, how many pencils will each student get?

Operations – Quiz

1. What is the place value of the Zero in 1,027?

2. What is the place value of the 7 in 7,002?

3. Write the standard number for 300 + 2 + 5:

4. Expand the number 7,329:

5. Write an equivalent equation for 3 tens + 26 ones:

6: If the time is 4:15 PM, what number is the hour hand pointing to?

7. If it is Tuesday, what is the name of the day to come next?

8. How many days per week do students go to school?

9. Since January is the first month of the year, what month is 4th ?

10. Fill in the spaces: 5 x _____ = 50 AND 50 ÷ 5 = _____

Money

It is very important to learn about money at an early age. In Second Grade students learn the correct value for each Coin and Bill. The money we earn and spend is referred to as Currency.

The list below contains the common coin and bill amounts that students will be expected to learn:

COINS	BILLS
Penny = 1¢	One Dollar = $1
Nickel = 5 Pennies = 5¢	Five Dollars = $5
Dime = 2 Nickels = 10 Pennies = 10¢	
Quarter = 5 Nickels = 25 Pennies = 25¢	

Dollars and Cents

Recognizing the types of Currency involved in a math problem is the first step to learning other functions involving money. The symbols shown next to the above coin and bill amounts represent dollars ($) and cents (¢). For the amounts below, indicate which are dollars and which are cents:

$5	3¢	7¢	$6	25¢	$17	$100	32¢	18¢	$3
Dollars	Cents	Cents	Dollars	Cents	Dollars	Dollars	Cents	Cents	Dollars

Indicate the type of coin or dollar that is represented below:

AMOUNT	CURRENCY
1¢	Penny
$1	Dollar Bill
5¢	Nickel
$5	Five Dollar Bill
10¢	Dime
25¢	Quarter

Similar to expanded numbers, dollars and cents can be grouped together in many different quantities and arrangements to create the same total amount.

For example, 15¢ can be separated into the following combinations of coins:

1 dime and 1 nickel = 10¢ + 5¢

1 dime and 5 pennies = 10¢ + 1¢ + 1¢ + 1¢ + 1¢ + 1¢

1 nickel and 10 pennies = 5¢ + 1¢ + 1¢ + 1¢ +1¢ + 1¢ + 1¢ + 1¢ + 1¢ + 1¢ + 1¢

Using the above concept create two different combinations of coins for the following amounts:

AMOUNT	COMBINATION 1	COMBINATION 2
30¢	2 dimes and 2 nickels	1 dime, 2 nickels, and 10 pennies
45¢	1 quarter and 2 dimes	4 dimes and 1 nickel
71¢	2 quarters, 2 dimes and 1 penny	1 quarter, 4 dimes, 1 nickel, and 1 penny
84¢	8 dimes and 4 pennies	3 quarters, 1 nickel, and 4 pennies
93¢	3 quarters, 1 dime, 1 nickel, and 3 pennies	9 dimes and 3 pennies

Dollar bills are made from the value from coins. Instead of carrying around hundreds of coins, certain amounts of money were printed on paper notes and used in place of the coins. The following amounts of coins equal one-dollar bill: 4 quarters, 10 dimes, 20 nickels, and 100 pennies. Now, create combinations of bills and coins from the following amounts:

AMOUNT	COMBINATION 1	COMBINATION 2
100¢	2 quarters and 5 dimes	3 quarters, 2 dimes and 1 nickel
137¢	1 dollar bill, 3 dimes, and 7 pennies	1 dollar bill, 1 quarter, 1 dime, and 2 pennies
150¢	1 dollar bill and 2 quarters	4 quarters and 5 dimes

Adding and subtracting monetary amounts is just like adding and subtracting regular numbers. Students will practice adding and subtracting money up to the amount of one dollar similar to the problems below:

```
  17¢     23¢     56¢     12¢     45¢     85¢     99¢     37¢     25¢     74¢
+  5¢   +  4¢   + 13¢   + 82¢   + 30¢   - 20¢   - 45¢   - 27¢   - 10¢   - 50¢
  22¢     27¢     69¢     94      95¢     65¢     54¢     10¢     15¢     24¢
```

To better represent how currency is written when dollars bills and coins are combined, the decimal point is used to separate coin amounts from dollar amounts. In addition, the coin amounts only go up to 2 places to the right of the decimal and only up to a total of 99¢. Once the amount of one dollar is reached, the amount is expressed on the left side of the decimal. An example of how dollars and cents are written together is presented below:

Dollars	Cents in amounts of 10¢	Cents in penny amounts (1¢)	Proper written total Amount
2	20	8	$2.28
1	50	7	$1.57
5	80	3	$5.83

As shown above, when dollars and cents are written as combined amount, only the dollar ($) symbol is used. Even if only cents are represented, the dollar sign is used as shown in the following example:

50 cents = 50¢ = $0.50

The zero to the left is a placeholder. The zero to the right of the 5 is actually the ones place (pennies) and is needed to complete the number Fifty (50).

The following examples using dollars and cents are written with the decimal point and dollar sign:

NOTE: unlike traditional decimal places, the first place is the tens and the second place is the ones.

Amount	Written Description	Amount	Written Description
$2.27	Two Dollars and Twenty-Seven Cents	$2.50	Two Dollars and Fifty Cents
$5.56	Five Dollars and Fifty-Six Cents	$1.25	One Dollar and Twenty-Five Cents
$1.75	One Dollar and Seventy-Five Cents	$4.80	Four Dollars and Eighty Cents
$8.32	Eight Dollars and Thirty-Two Cents	$3.90	Three Dollars and Ninety Cents

How many of the above amounts could have pennies included?

Five. The amounts with a number in the ones place ($0.01) are $2.27, $5.56, $1.75, $8.32, and $1.25.

How many of the amounts could have quarters in the cents portion?

All of them could have quarters. Since one quarter = 25 cents = $0.25, and the smallest amount above in the cents portion is $0.25, one or more quarters could be part of the total amount of cents in each money description.

Since exchanging money for goods is something we do every time we buy something, the following examples present situations involving adding and subtracting various amounts of money:

1. Sara has 3 quarters and 2 dimes (95¢). She wants to buy two candy bars for 20¢ each. Does she have enough money? Yes. 95¢ - 40¢ = 55¢

2. With her change (left over money) can she buy 2 more candy bars?

    Yes. 55¢ - 40¢ = 15¢

3. Based on her change in Question 2, how many nickels does Sara need to get one more candy bar?

    She needs a total of 20¢ to get another candy bar. She has 15¢, so she needs 1 nickel (5¢) to total 20¢

Money – Practice Sheet

Names and Amounts of Coins

1. Which coin has an amount of 25 cents?

2. Which coin has the amount of 1 cent?

3. What is the value of a dime?

4. What is the value of a nickel?

5. How many quarters in One Dollar?

6. How many dimes in One Dollar?

7. How many pennies in One Dollar?

8. How many nickels in one dime?

9. How many pennies in one dime?

10. How many dimes in one quarter?

Combinations and Amounts

1. How much money is 2 quarters and 3 pennies?

2. How much is 3 dimes, 2 nickels and 4 pennies?

3. How much is 3 quarters and 3 dimes?

4. How much is 2 Dollars and 4 dimes?

5. Which coins would be left if you started with 3 quarters and 3 dimes and spent 2 quarters and 1 dime?

6. If you started with 6 dimes and 3 nickels and spent 2 dimes and 1 nickel, how many quarters could you exchange the remaining dimes and nickels for?

7. If Tim started with $3.89 and bought a book for $1.23, what would he have left?

8. If Joy started with $3.25 and earned $2.00 for walking the dog and $1.00 for setting the table, how much will she have to go shopping?

9. How much money will Sam have at the end of the week if he earns $8.75 and spends $6.25?

10. If Josie starts with $9.85 and spends $7.38 on school supplies, how much would she have left?

Money – Quiz

1. How many dimes equal the same amount as 6 quarters?

2. What is the name of the coin that is equal to 5 pennies?

3. How many pennies in One Dollar?

4. How many nickels in 3 quarters?

5. What combination of quarters, dimes, and pennies is $0.98?

6. Write out the currency value for 3 quarters, 4 dimes, and 2 nickels:

7. If Danny earns $3.00 from his aunt, $2.50 from his Grandpa, and $1.25 from his cousin, what would be the total amount of money Danny earned?

8. If Lisa has 6 quarters and 21 pennies, does she have enough money to buy a slice of pizza for $1.75?

9. If Lisa borrows $0.20 from her friend Luke, what will her change be after buying the pizza?

10. How much money will Brian have if he finds 2 quarters, 6 dimes, 4 nickels and 13 pennies?

Shapes/Patterns

Many patterns are found throughout the subject of Second Grade Math. There are repeating patterns, increasing patterns, decreasing patterns, and finally counting patterns.

Patterns are seen all around us. For example, traffic lights present a repeating pattern. They turn green to indicate to drivers that they should move forward through the intersection, second, they turn yellow to warn drivers to slow down because finally, they turn red to alert cars to stop. This pattern is repeated many times daily to prevent accidents within the intersection. An example of an increasing pattern is the scoring used in football games. Each touchdown is 6 points (I am not counting the extra point to keep this simple). So, each time the team scores, their total points increase by 6. Basically, it is counting by sixes.

Repeating Patterns

Before learning how to use counting patterns, Second Grade students learn how to identify patterns by using shapes to represent the items in the pattern. The following examples represent repeating patterns:

1. Based on the above pattern, which picture will be the next in line? Moon

2. Write in words what this pattern is repeating.

    Answer: star, moon, star, moon, star, moon, star, moon, star

3. If a second moon is added to each current moon, what would the pattern look like?

   Answer: star, moon, moon, star, moon, moon, star, moon, moon, star, moon, moon, star

4. What would the pattern be if a sun were added after each moon?

   Star, moon, sun, star, moon, sun, star, moon, sun, star, moon, sun, star

Repeating patterns can involve many different elements. For example, instead of just repeating two single items, the pattern can repeat as many groups of items as you can possibly think of, but that gets a bit complicated for Second Grade Math. Another simple repeating pattern could be using pairs of items, instead of just single items. If that were incorporated into the above pattern concept, it would look like this:

Increasing Patterns

Items that are arranged in an increasing pattern can be very simple up to being quite complicated. Increasing patterns involve adding at least one item to one element of the pattern. More than one item can be added to the element and more than one element in a pattern can be increased. The following pattern only has one element increasing throughout:

1. Which shape in the pattern is increasing in number? Square

Many times, a single element will be chosen to be the item that is increased according to a certain formula for the pattern. The above pattern shows that the square is increased by one each time it follows a circle.

2. Given the pattern, how many squares will be added after the next circle?

Based on the pattern of adding one square to each set of squares after each circle, the next set of squares should have 5 items.

3. Do the circles change in this pattern?

No. The circles are not part of the increasing element.

Decreasing Pattern

This type of pattern is similar in concept to the increasing pattern, except it is more like a countdown. An element in the pattern begins with a larger quantity and gets smaller as the pattern progresses. The example above with the squares and circles can be turned into a decreasing pattern by going in reverse order. The pattern would then be 4 squares, 1 circle, 3 squares, 1 circle, 2 squares, 1 circle, and finally, 1 square.

Counting and Number Patterns

Using the above examples, determine whether the following number patterns are repeating, increasing, or decreasing:

A.   1 2 1 2 2 1 2 2 2 1 2 2 2 2 Increasing Pattern

B.   3 3 5 5 3 3 5 5 3 3 5 5 3 3 Repeating Pattern

C.   100 99 98 97 96 95 94 Decreasing Pattern

Other Number/Counting patterns include the "skip counting" students learn to help them memorize certain addition and later multiplication problems.

The beginning of "skip counting" usually starts with counting by twos. It will be apparent which type of pattern "skip counting" produces when seen below:

Twos   2 4 6 8 10 12 14 16 18

Threes   3 6 9 12 15 18 21 24 27

Fours   4 8 12 16 20 24 28 32 36

Fives   5 10 15 20 25 30 35 40

As can be seen from the above "skip counting" patterns, they are an increasing type pattern.

Using other types of numbers and number sequences can produce other types of patterns. For example, the pattern below uses even and odd numbers, what type of pattern is it?

3 4 3 4 3 4 3 4 3 4   This produces a Repeating Pattern

5 10 15 20 25 30 35 40   Counting by 5s produces an Increasing Pattern and a
  Repeating Pattern of Odds and Evens

The following examples show how patterns can be created by implementing one or more basic rules for the specific math operation:

1. Add two to every even number when counting by Ones: 1, 4, 3, 6, 5, 8, 7, 10, 9, 12, 11

2. Count by 3s from 23: 23, 26, 29, 32, 35, 38, 41, 44, 47, 50

3. Count by 5s from 50: 50, 55, 60, 65, 70

Fill in the Missing Numbers

1. 12, 14, ____, 18, 20, 22   The pattern is counting by 2s, so the answer is 16.

2. 40, 50, 60, ____, 80, 90   The pattern is counting by 10s, so the answer is 70.

3. 93, 90, 87, ____, 81, 78   The pattern is decreasing by 3s, so the answer is 84.

What is the next number in the Sequence?

1. 1, 2, 1, 2, 1, 2, 1, ____ The pattern is repeating 1 and 2, so the answer is 2.

2. 3, 1, 1, 4, 1, 1, 5, 1, 1, ____ The pattern is increasing by 1s after two consecutive 1s, so the answer is 6.

3. (1+2), (1+3), (1+4), ____ The pattern is showing adding 1 to a single number in an increasing pattern, so the answer is (1+5).

Shapes/Patterns - Practice Sheet

Define the following types of Patterns

1. ■ ■ ▲ ▲ ■ ■ ▲ ▲

2. ● ● ● ● ◆ ● ● ● ◆

3. ★ (★ ★ (★ ★ ★ (

Based on the above patterns, Answer the following questions:

1. What is the next shape in the first pattern?

2. What is the next shape in the second pattern?

3. What is the next shape in the third pattern?

Fill in the Blanks

1. 5, 10, _____, 20, 25, 30   2. 2, 5, 8, ____, 14, 17, 20, 23

3. 2, 4, 6, 20, _____, 60, 200, 400, 600, 2,000, 4,000

4. 100,000, 10,000, _____, 100, 10, 1

Find the next number in each sequence

1. 1, 3, 5, 7, _____   2. 3, 6, 9, 12, 15, _____

3. 2, 4, 6, 8, _____   4. 35, 40, 45, 50, _____

Alternate numbers with the letters of the alphabet from A to Z

1. Count by 2's

2. Count by 3's

3. Count by 5's

4. Count by 10's

True or False

1. Counting by 2s represents an increasing pattern.

2. A countdown is considered to be a repeating pattern.

Shapes/Patterns Quiz

1.

How many squares are repeated in the pattern?

How many triangles are repeated in the pattern?

2.

How many circles repeat in this pattern?

How many diamond shapes repeat?

Complete the following patterns:

1. 1, 3, 4, 7, 11, _____

2. 2, 4, 6, 12, 14, 16, 22, 24 _____

3. 81, 72, 63, 54, _____

Fill in the blanks:

1.

2. 5, 10, 15, _____, 25, _____, 35, 40

3. 3, 7, 10, 17, 27, _____

True or False

1. If the following pattern continues, the next shape will be rectangle.

    Rectangle, square, triangle, rectangle, square, triangle, rectangle, _____

2. Counting by fives produces two patterns: an increasing pattern and a repeating odd and even pattern.

Length

The standard system of measuring length in the United States uses inches, feet, yards, and miles. Based on the length or distance being measured, a different unit is chosen.

For example, to describe how far one city is from another, it would be more appropriate to use miles rather than inches. However, inches would be more appropriate for measuring the length of a small table. Feet are often used to measure a person's height and yards are commonly used for measuring plots of land like football fields, which are 100 yards long.

Length Equivalents

Similar to other types of measurement, the different lengths have equivalent amounts to each other. The following list shows the various Standard Unit Lengths and how they relate to one another:

12 Inches = 1 Foot

3 Feet = 1 Yard   36 Inches = 1 Yard

1,760 Yards = 1 Mile   5,280 Feet = 1 Mile

Based on the above information, answer the following questions with the appropriate measurement:

1. To figure out the length of ribbon needed to wrap a birthday present, do you use   yards or inches?

Inches are a better unit of measurement because the amount of ribbon needed is closer to inches than yards.

2. To measure the size of a living room floor to see if a special rug will fit, is it better to use miles or feet?

Feet are a more appropriate unit of measurement for a room. Miles are used to measure larger lengths and distances.

3. To calculate the distance between Houston, Texas and New York City, New York, is it better to use feet or miles?

Miles are a better unit of measurement because the length or distance from Houston to New York is much greater than the size of a foot.

4   To figure out the length of a strip of land to determine if a new football field can be installed, are inches or yards a better unit of measurement?

Yards are a better unit of measurement because the length of a football field is generally measured in 10-yard sections up to a total of 100 yards.

In school, students practice measuring items using a ruler with 1-inch markings up to 12 inches. The inches are also divided in half and have marking associated with them so students can measure items smaller than 1 inch. Answer the following measurement questions based on the use of a standard 12-

inch ruler:

1. Jamie measures two pencils using his ruler. One measures 3 inches and the other measures 2 and 1/2 inches. Which one is shorter in length?

The pencil that is 2 and 1/2 inches is shorter.

2. Which is closer to a foot in length: 7 inches or 13 inches?

13 inches, although greater than one foot, is only 1 inch from 12 inches (1 foot) and 7 inches is 5 inches away from a foot. So, 13 inches is closer to 1 foot than 7 inches.

3. One book measures 8 and 1/2 inches wide and another measures 9 inches across. Which book has a larger width?

The book that measures 9 inches across has a larger width than the book that measures 8 and 1/2 inches.

4. How many 12-inch rulers would a student need to measure an item that was 18 inches long?

One whole 12-inch ruler and 6 more inches of a second ruler would be needed to measure an item 18 inches long.

Standard Length Units and Metric Length Units

The student would need two rulers and would measure the entire 12 inches of one and then 6 inches of the second to be able to accurately measure the item.

Although in the United States we mainly use the Standard Units of Length to measure items and distances, the Metric System is used on occasion and is used frequently throughout the world. The Metric System uses millimeters, centimeters, meters, and kilometers to measure lengths and distances.

Below is a list of general equivalents between the Standard Units of Length and the Metric Units of Length:

1 meter is approximately 2 1/2 feet

1 mile is approximately 2 kilometers

1 inch is approximately 2 1/2 centimeters

Based on the their approximate equivalents, answer the following questions comparing Standard Units and Metric Units:

1. Which is greater: 1 Meter or 1 Inch?   If 1 Meter is approximately 2 1/2 feet and it takes   12 inches to equal 1 foot, 1 Meter is greater than 1 Inch.

2. If miles are used to measure the distance between cities, which is the similar Metric Unit: Centimeters or Kilometers? Since, 1 mile is approximately equal to 2 Kilometers and 2 1/2 Centimeters is approximately 1 Inch, the similar metric unit to Miles is Kilometers.

3. If it is better to measure a room in feet, which Metric Unit would also be used to measure the length of a room? Since 2 1/2 feet equal 1 meter, the Metric Unit to measure the length of a room would be the Meter.

4. If a town is 12 miles from the nearest grocery store, approximately how many kilometers from the grocery store is the town? Since, 1 mile is approximately 2 kilometers, the grocery store and town are about 24 kilometers apart.

5. A student measures himself to be 5 feet tall. Approximately, how many meters tall is the student? Since there approximately 2 1/2 feet per meter, the student is about 2 meters tall.

In summary, whether using Standard Units or Metric Units, the following concepts need to be remembered by Second Grade students:

- Inches and Centimeters are the best units to measure the length of smaller items like books, pencils, ribbon, shoe size, etc.
- Feet and Meters are best for measuring the length moderately sized objects and items like the length of a room, the size of a ping-pong table, the proper length for curtains, etc.
- Miles and Kilometers are best for measuring distances between neighbors, towns, and cities.
- Generally, heights are measured in feet and meters.

Length – Practice Sheet

1 meter is approximately 2 1/2 feet 1 mile is approximately 2 kilometers 1 inch is approximately 2 1/2 centimeters	12 Inches = 1 Foot 3 Feet = 1 Yard 1,760 Yards = 1 Mile	36 Inches = 1 Yard 5,280 Feet = 1 Mile

Standard Length Units

1. Which is greater, 12 inches or 2 feet?

2. Which is greater, 3 feet or 3 yards?

3. Which is smaller, 12 inches or 1 foot?

4. A Person's height is usually measured in _____.

5. The height of a mountain is usually measured in _____.

6. Which is greater 12 yards or 1 mile?

7. Would a student measure his height using inches or miles?

8. Which is a better unit of measure for determining the distance between two cities,   inches   or   miles?

9. How many feet are in 1 yard?

10. How many feet in 1 mile?

Metric Length Units

1. Which is longer, 1 meter or 1 centimeter?

2. which is better for measuring distances between cities, centimeters or kilometers?

3. Which is better for measuring the length of a spoon, centimeters or meters?

4. Which is better for measuring the length of a driveway, kilometers or meters?

Comparing Standard and Metric Length Units

1. Which is greater, inches or meters?

2. Which is smaller, feet or kilometers?

3. Which Standard and Which Metric Units are best for measuring the distance between San Diego and Los Angeles?

4. Approximately, how many centimeters are in 1 inch?

5. Approximately, how many kilometers in 1 mile?

6. Approximately how many feet in 1 meter?

Length – Quiz

1. Which Standard Unit is better for measuring the distance between two windows of a house?

2. Which Standard Unit is best for measuring the length of fabric needed to re-cover several pieces of furniture?

3. Which Standard Unit is best for determining how far San Francisco is from New York City?

4. Which Standard Unit is generally used for measuring a person's height?

5. How many inches in 1 foot?

6. How many feet in 1 yard?

7. Which are similar in length, miles and centimeters or meters and feet?

8. Which are used for measuring long distances, miles and kilometers or inches and centimeters?

9. Approximately, how many kilometers in 1 mile?

10. Approximately, how many centimeters are in 1 inch?

Weight/Capacity

Measuring the weight and capacity of objects and items is a daily occurrence in most households. For example, most mornings, people weigh themselves before dressing for work or school, people with special dietary needs weigh their food to be sure they are getting the proper amount of specific nutrients, the volume of water poured into a coffee maker has to match the capacity of the machine so it does not overflow, and the capacity of a tent needs to be known before heading out on a camping trip.

Weight is the measurement of how heavy something is and capacity is the measurement of the inside of an object to determine how much "stuff" it can hold. Both weight and capacity are measured in the United States by using Standard Units of Weight and Standard Units of Volume. The Standard Units for Weight are ounces, pounds, and tons. The Standard Units for Capacity are cups, pints, quarts and gallons.

Standard Units of Weight

The units of measurement for weight are based on the size and anticipated weight of an object or item. The material an object or item is made of also determines which unit of measure will be used to describe the weight.

For example, if a bag of feathers and the same size bag of concrete bricks are going to be weighed, which one is expected to be heavier? Well, based on the experience of feathers being much lighter than concrete bricks, it is anticipated that the bag full of bricks will be heavier.

The table below lists the Standard Units of Weight and common items those units measure:

OUNCES	Individual Fruit and Vegetables, flour and sugar for baking, baby animals, sliced cheese and deli meats, wool and yarn.
POUNDS	Bunches of fruits and vegetables, turkeys, potatoes, bundles of firewood, bags of grain, people, backpacks, books, coffee.
TONS	Cars and trucks, elephants, logs, piles of bricks,

Weight Comparisons and Equivalents

In Standard Units of Weight, the order of smallest to greatest unit of measure is ounces, pounds, and tons: ounces being the smallest unit of measure and tons being the largest. Since pounds fall somewhere in between ounces and tons, it is the most popular unit of measure.

Based on the above information, answer the following questions:

1. Which unit would be used to measure the weight of an apple? Since apples are relatively small and light, they would be weighed using ounces.

2. Which unit would be used to measure a truck carrying a load of bricks to a construction site? Since a truck carrying bricks is considered to be very heavy, it would be weighed using tons.

3. Is a golf ball more likely to weigh 2 ounces or 2 pounds? Since a golf ball is relatively light and needs to be aerodynamic, it is more likely to weigh 2 ounces.

4. Is a bicycle more likely to weigh 15 pounds or 15 tons? Since a bicycle is a moderately-sized item, it is more likely to weigh 15 pounds.

To convert between ounces and pounds, and between pounds and tons, refer to the following information:

16 ounces (oz) = 1 pound (lb)

2,000 lbs = 1 Ton (T)

Convert between Standard Units of Weight

1. How many ounces of ice cream are needed to make 2 lbs? Since it takes 16 oz to make 1 lb, it will require 32 oz to make 2 lbs of ice cream.

2. How many pounds does a 3-Ton boulder weigh? Since there are 2,000 lbs in 1T, there are 6,000lbs in 3T.

Metric Units of Weight (Mass)

Although Standard Units of Weight are the units of choice most of the time in the United States, sometimes Metric Units of Weight (Mass) are used. So, it is important to be introduced to Metric Units of Weight early in a student's education.

The following Metric Units and their similar Standard Units of Weight have been matched:

2 pounds are approximately 1 kilogram

1 ounce equals approximately 30 grams

Generally, pounds and kilograms are the most widely used units of weight measurement. Based on the above conversion information, indicate approximate equivalent weights for the items described below:

1. How many pounds is a table that weighs 5 kilograms? Since each kilogram is equal to about 2 pounds, the approximate weight of the table is 10 pounds.

2. How many ounces is 90 grams? Since there is 1 oz for each 30 grams, then 90 grams is approximately 3oz.

3. Approximately how many kilograms does a 50-pound desk weigh? Since there is one kilogram per 2 lbs, the desk weighs about 25 kilograms.

Standard Units for Capacity

Capacity is the amount of a substance in a container. These containers have the following Standard Capacities from smallest to largest: cups, pints, quarts, and gallons. These units are used to measure numerous amounts of substances daily in most households. For example, a typical breakfast may consist of a cup of milk with a bowl of cereal, to make chocolate chip cookies the ingredients are usually measured using cups, and when milk or juice is purchased at the store, they are usually packaged in gallon and 1/2 gallon containers.

Comparing Units of Capacity

For the following items, decide which unit of measurement is best:

1. Is the capacity of a swimming pool better measured in gallons or pints? Since a pool can hold a very large amount of water, gallons are the best unit of measurement.

2. When adding water to a small fish tank, would it be better to add a few cups or several quarts? Since the fish tank is described as being small, using cups to add water would be the best choice.

3. A mom is making hot chocolate for her two children. Is it better for her to use cups to measure the amount of hot chocolate per child or gallons? Since the mom is making hot chocolate to be consumed by just two children and not a large group of people, cups is the better unit of measure.

Sometimes, these units of measure need to be converted one to another and the following list indicates the proper conversions:

2 cups = 1 pint

2 pints = 1 quart

4 quarts = 1 gallon

Calculate the proper amounts in the following problems:

1. If a recipe calls for 2 quarts of milk, how many pints will the recipe need? Since 2 pints equal 1 quart, then 4 pints equal 2 quarts and 4 pints would be needed for the recipe.

2. Water needs to be available to the competitors of a foot race. If there are 16 cups in a gallon, how many gallons will be needed to produce at least 64 cups? If 16 cups make a gallon and 64 cups can be divided into 4 groups of 16 cups, then 4 gallons of water will be needed for the race.

3. If 3 gallons of paint is used to cover the inside walls of 3 bedrooms, how many quarts of paint does that amount equal? Since there are 4 quarts per gallon, then 3 gallons yields 12 quarts.

Metric Units for Capacity

The most popular unit of measurement in the Metric System is the Liter (l). The closest Standard Unit of Measurement is the gallon. One gallon equals approximately 2 liters.

Understanding the approximate equivalent of 2 liters to 1 gallon, solve the following conversion problems:

1. Several people will be attending a party. If the hosts have a 5-gallon container from which soda-pop will be dispensed, how many 2-liter bottles of soda-pop will they need? Since 1 gallon is approximately equal to 1 2-liter container, 5 2-liter bottles will need to be purchased.

2. Each student going on a field trip needs to bring a 1-liter container of water to drink. How many gallons will the teacher need to fill 20 liters? Since there are approximately 2 liters per gallon, the teacher will need to use 10 gallons of water.

Summary

Whether students use Standard Units of Measurement for weight and capacity or Metric Units of Measurement for weight and capacity, understanding when to use which unit is the most important concept.

In other words, smaller units for smaller, lighter items and objects, and larger units for larger items and objects.

Standard Units for Weight smallest to biggest = ounces, pounds, tons

Metric Units for Weight smallest to biggest = grams, kilograms

Standard Units for Capacity smallest to biggest = cups, pints, quarts, gallons

Metric Unit used most often for Capacity = liter

Weight/Capacity – Practice Sheet

1. A fully loaded logging truck stops to be weighed. Will its weight be measured in tons or pounds?

2. Several students weigh themselves for a science experiment. Should their weight be recorded in pounds or ounces?

3. A pizza parlor weighs the amount of shredded cheese it uses for each pizza. Do they use ounces of cheese or pounds of cheese per pizza?

4. What is the order of the Standard Units for Weight from lightest to heaviest?

16 ounces (oz) = 1 pound (lb)

2,000 lbs = 1 Ton (T)

5. How many ounces are in 2 pounds?

6. How many Tons will 6,000 pounds create?

7. How many pounds will 48 ounces create?

2 pounds are approximately 1 kilogram

32 ounces are approximately 1 kilogram

1 ounce equals approximately 30 grams

8. If a great big science book weighs 4 pounds, approximately how many kilograms does it

weigh?

9. 2 kilograms equals approximately how many ounces?

10. A 2-ounce golf ball weighs approximately how many grams?

11. Which units of weight (standard and metric) would be best for measuring how much a ball of yarn weighs?

2 cups = 1 pint

2 pints = 1 quart

4 quarts = 1 gallon

12. If 6 pints of cream are needed for a pie recipe, how many quarts should be used?

13. 20 gallons of punch are needed for a party. How many quarts can be used?

14. How many cups will 15 pints of apple juice produce?

15. How many pints equal one gallon?

16. Which is the best unit of measure to determine the capacity of a swimming pool?

17. Which unit of capacity is best for measuring how much sugar is needed for a cake recipe?

18. Is it more likely to find quarts of juice at the store or cups?

19. How many liters equal approximately 1 gallon?

20. Are gallons or liters of soda-pop sold in the average grocery store?

Weight/Capacity – Quiz

16 ounces (oz) = 1 pound (lb) 2,000 lbs = 1 Ton (T)	2 pounds are approximately 1 kilogram 32 ounces are approximately 1 kilogram 1 ounce equals approximately 30 grams 1 gallon is approximately 2 Liters	2 cups = 1 pint 2 pints = 1 quart 4 quarts = 1 gallon

1. Which of the above units represent the Standard Units for Weight?

2. How many Liters approximately equal 1 gallon?

3. In an Ice Cream eating contest the winner ate a total of 8 pints. How many quarts did the winner eat?

4. Twelve cups of pineapple are needed for a special recipe. How many 1-quart cans of pineapple are required?

5. Approximately, how many liters of water will be needed to fill 8 1-gallon containers?

6. A wheelbarrow carries a small pile of bricks to the backyard of a home to build a new fire pit. Should the bricks be weighed using pounds or ounces?

7. Approximately how many pounds does a 5-kilogram package weigh?

8. How many cups are needed for a recipe that calls for 6 pints of cream?

9. A huge slap of stone that weighs 3 tons is going to be carved into a statue. The artist has to pay for the stone by the pound. How many pounds does he have to pay for?

10. If an item weighs 2 kilograms, approximately how many ounces does it weigh?

Probability

Probability is the study of "what if?" In other words, if a specific scenario is presented, what is likely or unlikely to occur when certain elements are introduced. Or, what is the likelihood of something happening based on a specific set of circumstances.

For example, is it more likely or less likely that you will see a giraffe walking down the middle of the street in New York City? Since giraffes are not native to New York City, the fact that New York City is not located in Africa, and if, the zoo in New York City did house giraffes, it would be almost impossible for them to escape, it is less likely that you will see a giraffe walking down the middle of the street in New York City.

Likely, Unlikely, Equally Likely

In Second Grade students are presented with the concept of probability by assessing whether an occurrence of an event is likely to happen or unlikely to happen, and if there is a choice between two outcomes of a test, whether one outcome is more likely, less likely, or if both outcomes are equally likely. First, the concept of the probability of an event occurring during specific circumstances is learned.

The following examples express the concepts of likely or unlikely:

1. The sun is shining and the sky is a beautiful clear blue. Is it likely or unlikely to rain?

It is unlikely to rain because the description of the day indicates no clouds and a lot of bright sunshine.

2. It is a very hot day and the power has gone out in a neighborhood that has a large neighborhood pool. Is it likely or unlikely that all the neighbors will go swimming?

It is very likely the pool will be crowded because the pool would be a great place to stay cool if the weather is very hot.

3. The students of a second grade math class have studied very hard for an upcoming test. All of them pass and get above 80% on their pre-tests. It is likely or unlikely the majority students will pass the test with at least 80% correct?

It is likely that the majority of the students will pass with at least 80% because of the positive results of the pre-test.

4. A certain baseball player hits a homerun 2 times for each 10 times at bat. It is likely or unlikely that on any given pitch, this baseball player will get a homerun?

It is unlikely because he only gets a homerun a couple times for every 10 times he is at-bat.

In addition to determining if a single event is likely or unlikely, students will need to assess the probability of which one of two possible outcomes is more likely, less likely, or if both outcomes are equally likely. The following examples illustrate this concept of less likely, more likely, or equally likely:

1. Two basketball teams are playing the final game of the season. Whichever team wins will get the first place trophy. If team A has 4 players over 6 feet tall and team B has 1 player over 6 feet tall, which team is more likely to win? If the players are all equally skilled, it is more likely team A will win because more of their players can reach the basket and block the ball more easily due to their height advantage.

2. A dartboard is covered in the color blue with the exception of a very small spot of red in the middle. On which color is a dart more likely to land? A dart is more likely to land on the blue color because it covers a much greater area of the dartboard.

3. If a bag contains 4 blue marbles and 4 red marbles, how likely is it that a blue marble will be blindly chosen? Since the number of blue marbles and the number of red marbles is the same, there is an equal chance that either color will be chosen.

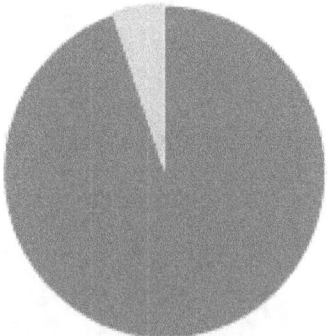

4. If paintballs were shot at the above target, would they be more likely to hit the light gray area or the dark gray area? Since the dark gray area is greater in size and easier to hit, it is more likely that the paintballs will hit the dark gray area than the light gray area.

Certain, Probable, Unlikely, Impossible

Based on certain information, students will be able to determine whether an event is certain, probable, unlikely, or impossible. For example, if a bed of flowers is filled completely with only red flowers, the likelihood of finding a white flower is Impossible. Conversely, the likelihood of finding a red flower in the flowerbed is certain.

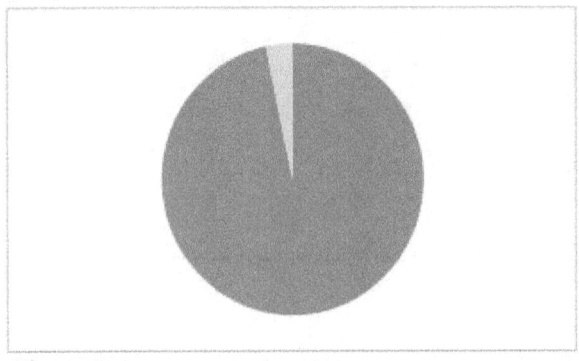

Chart A

Chart B

Based Chart A, is it probable or unlikely that an arrow would strike the light gray area? Since the light gray area is significantly smaller than the dark gray area it makes it a much more difficult area to hit. Therefore, it is unlikely that an arrow will strike the light gray area. Because the light gray area exists within the dark gray area, the possibility exists that it could be struck so it cannot be considered impossible. Similarly, it cannot be considered certain that the dark gray area will be struck.

Although similar to Chart A, Chart B represents a set of probabilities that are both impossible and certain. An arrow shot at this target will certainly hit the gray area. The likelihood that an arrow will strike a different color is therefore, impossible.

Other situations that involve impossible and certain probabilities occur every day. Examples include the following:

1. It is certain that the sun will not be seen at night.

2. It is impossible for a dog to have a litter of kittens.

3. It is certain that a fire in the fireplace is extremely hot.

4. It is impossible for ice to catch fire.

5. It is certain that the earth has only one moon.

6. It is impossible for humans to breathe naturally underwater.

Students will use math and probability equations to determine the probable occurrence of specific events with better accuracy later in elementary school and middle school.

Probability – Practice Sheet

True or False

1. If an event is considered to be likely to occur, then it will definitely occur every time the test is conducted.

2. If an event is considered to be unlikely, then although it will not occur often, the possibility that the event could occur does exist.

3. When an event is certain to occur, there is still a small chance it will not occur on occasion.

4. When an event is considered to be impossible, it will never occur as long as the circumstances remain the same.

More Likely, Less Likely, Equally Likely

1. A spinner used for playing board games is half red and half blue. What is the likelihood that the arrow will land in the blue area?

2. During the winter months in Michigan, what is the likelihood that it will snow?

3. During the hot summer months in the desert, what is the likelihood of rain?

4. A bag full of jellybeans has 100 red jellybeans and 5 yellow jellybeans. Which color is more likely to be grabbed in a handful?

5. A deck of cards has exactly 4 kings and 4 queens. If only these 8 cards are used, what is the likelihood of drawing a king rather than a queen?

6. A teacher hands out pencils to her class to take a test. There are forty students and 50 pencils. If only two of the pencils are sharpened, what is the likelihood that a student will get a sharpened pencil?

7. A tennis champion has won 9 out of the last 10 matches against his opponent. Is it more likely or less likely that he will win the next match against this same opponent?

8. If a second grade student reaches into a bag containing the last 10 chocolate bars and last 3 lollipops from Halloween, which sweet treat is more likely to be grabbed?

9. Mr. Smith has carved the Smith Family Thanksgiving turkey for the last 8 Thanksgiving dinners. It is more likely or less likely that he will carve the turkey again this year?

10. Madison watches cartoon everyday after school for a half-hour before doing her homework. Is it more or less likely she will do her homework before watching cartoons tomorrow after school?

Certain, Probable, Unlikely, Impossible

1. The sun rises in the east and sets in the west. What is the likelihood that it will rise in the west

tomorrow morning?

2. A robin lays three eggs in the spring. What is the likelihood that all three chicks will be baby robins?

3. A monarch caterpillar spins a cocoon and undergoes the change into a butterfly.   What is the likelihood that it will emerge as a Blue Morpho butterfly?

4. A bag contains 25 black checkers and 1 red checker. What is the likelihood that a   black checker will be blindly chosen?

5. A dozen chocolate chip cookies are made with semi-sweet chocolate chips. What   is the likelihood that there will be a cookie with white chocolate chips?

6. A package of cherry popsicles is purchased at the store. What is the likelihood   that when opened the package will contain grape popsicles?

Probability – Quiz

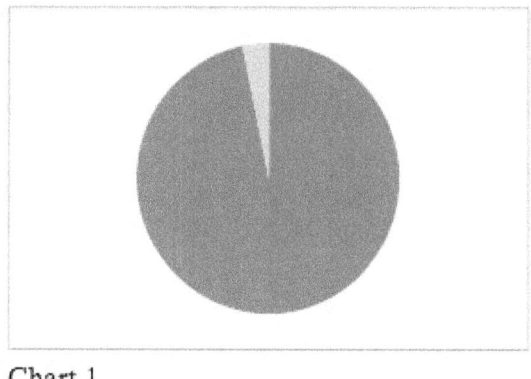

Chart 1   Chart 2

1. Which graph shows the possibility of two event outcomes?

2. Which graph shows either a certain outcome or an impossible outcome?

3. In Chart 1, which color depicts the less likely outcome?

4. What is the likelihood of an arrow hitting a white slice on Graph 2?

5. What is the likelihood of a paintball striking the dark gray area on Chart 1?

6. If the light gray portion of Chart 1 was increased to cover half of the circle, would the likelihood of it being struck by a paintball increase or decrease?

7. A box of crayons contains 20 purple and 20 orange colored crayons. If a student reaches into the box without looking into it, what is the likelihood that an orange crayon will be chosen opposed

to a purple crayon?

8. A teacher passes out a spelling test. All the students passed the pre-test given the day before. Is it more or less likely that all of the students will pass the test.?

9. A newspaper boy always throws the paper so it lands on the front porch of a specific house. When he delivers his paper tomorrow morning, what is the likelihood that the paper he throws lands on the front porch again?

10. As the earth circles the sun, seasons change in certain parts of the world. The cycle is winter, spring, summer, and fall. What is the possibility that summer will follow winter next year?

Graphs

Data is constantly being collected and interpreted throughout our modern society. Each night, if we watch the news programs, the weather report is usually presented. The meteorologists measure and interpret numerous groups of data to assess what the upcoming days will be like with regard to temperature, wind speed, cloud cover, rain, etc. The weather map we see on the news is essentially a weather graph.

Students in second grade will begin to learn how to read a variety of types of graphs and interpret the data presented. The types of graphs students will learn and use include coordinate graphs, bar graphs, pictographs, line plots, and line graphs.

Coordinate Graphs

The graph presented below is known as a Coordinate Graph. A vertical number and a horizontal number are paired into what are called coordinates. These coordinates are then plotted on a graph as a single location.

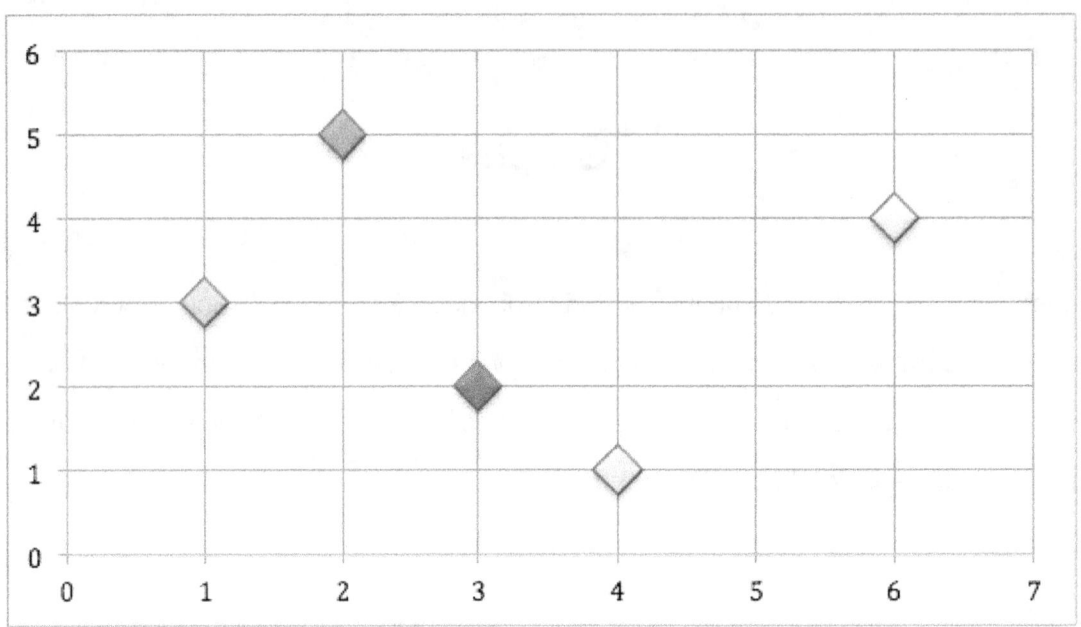

The Coordinate Graph above has 5 points plotted in a variety of locations. The following coordinates and their respective colored points are represented:

Descriptive Coordinate Location	Proper Coordinate Location
Violet is located at 1 across and 3 up.	Violet is at (1,3)
Red is located at 2 across and 5 up.	Red is at (2,5)
Blue is located at 3 across and 2 up.	Blue is at (3,2)
Green is located at 4 across and 1 up.	Green is at (4,1)
Yellow is located at 6 across and 4 up.	Yellow is at (6,4)

The first number in the proper coordinate is the horizontal number and the second number is for the vertical number.

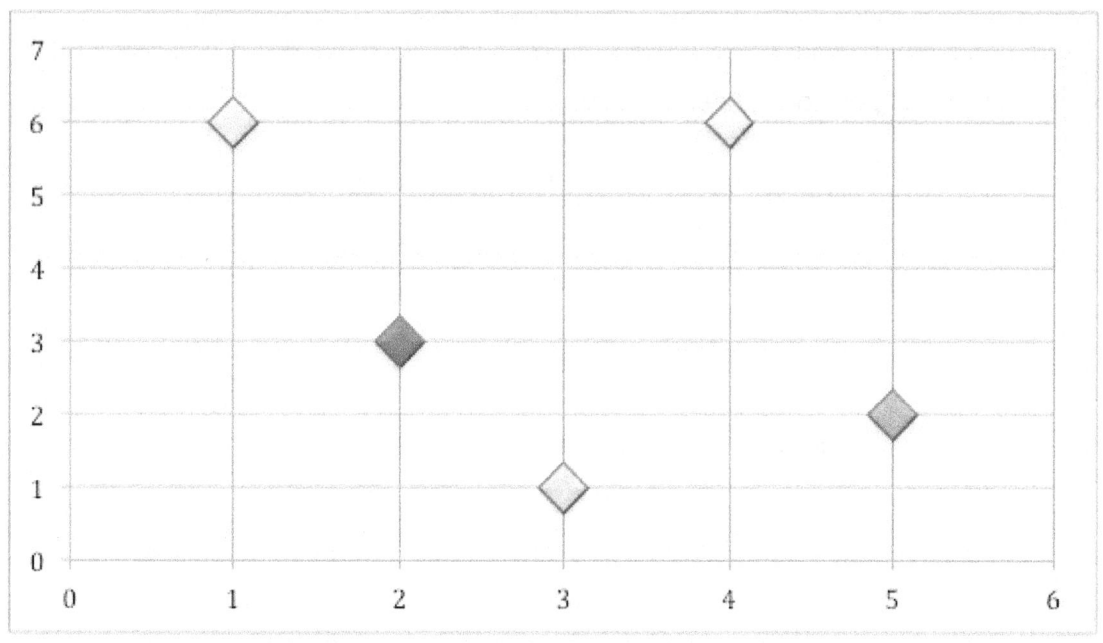

For the above graph, pair up the color of the coordinate with its horizontal and vertical numbers:

Green (1,6)   Blue   (2,3)   Violet (3,1)   Yellow (4,6)   Red (5,2)

By looking at the locations of each of these colored points, the student can make some interpretations:

1. The yellow and green points are the closest to the top of the graph.

2. The violet point is located closest to the bottom of the graph.

3. The red point is the farthest to the right on the graph.

4. The green point and the red point are farthest apart.

5. The blue point and the violet point are closest together.

6. The horizontal coordinate for the green, violet and red points are odd.

7. The yellow point is the only coordinate with two even numbers.

Bar Graphs

Another type of graph students will use is the Bar Graph. This type of graph is very easy to read because it displays collected data very clearly. The taller the bar, the bigger the number or the greater the quantity of data.

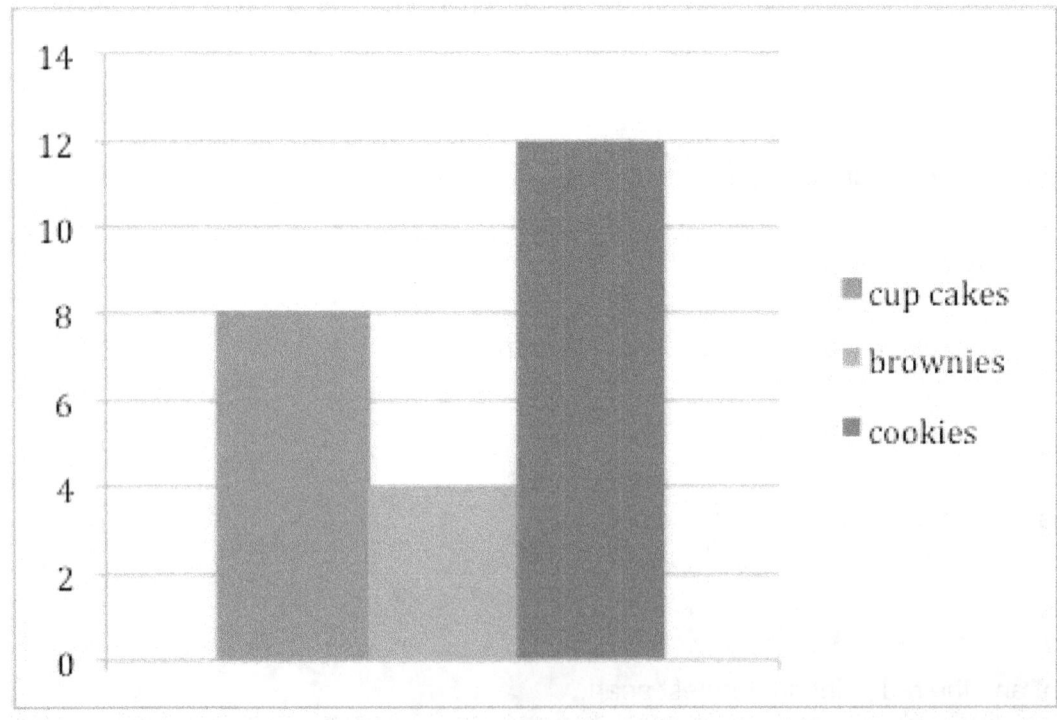

The above graph was created from the following data:

Item	Number Sold
Cupcakes	8
Brownies	4
Cookies	12

Information that can be gathered from this table and graph combination includes the following:

1. The greatest number of goods sold are the cookies.

2. The fewest goods sold are the brownies.

3. Four more cupcakes were sold than brownies and four more cookies were sold than cupcakes.

4. Cookies were the most popular item and brownies were the least popular item.

Based o the above information, some conclusions can be approximated:

1. More people like cookies than brownies and cupcakes.

2. The cookies may have been less expensive than the cupcakes and brownies.

3. The cupcakes and brownies may have been larger and were shared by a couple people and the cookies were smaller and enjoyed by individuals.

Pictographs

Pictographs are another form of presenting data in an easily understandable visual format. Items are used in a table format to represent certain quantities of that item.

Betty	🥚 🥚 🥚
Lucy	🥚
Sally	🥚 🥚 🥚 🥚 🥚
Amy	🥚 🥚
Kathy	🥚 🥚 🥚

The above pictograph represents 5 hens and the number of eggs they laid over 5 days.

Based on this graph, the following information can be collected:

1. Sally laid one egg each day.

2. Lucy only laid one egg over 5 days.

3. Amy laid one more egg than Lucy.

4. Betty and Kathy laid equal numbers of eggs.

5. Sally is the best producer of eggs in this hen house.

6. The total number of eggs produced in this hen house is 14.

Line Plots

Another graphing method that could be used for the above data with a couple of details added is called a Line Plot. The following Line Plot shows how many eggs per day were laid by the hens.

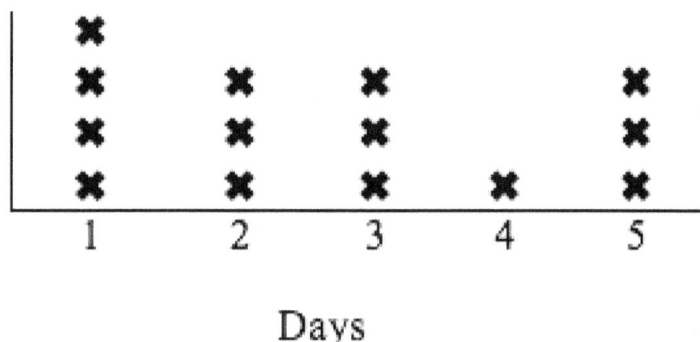

This graph shows that only Lucy laid any eggs on Day 4, but that at least two other hens laid an egg on each of the other five days.

Line Graphs

The last type of graph to be covered is called a Line Graph. This graph shows several data points connected to form a continuous line.

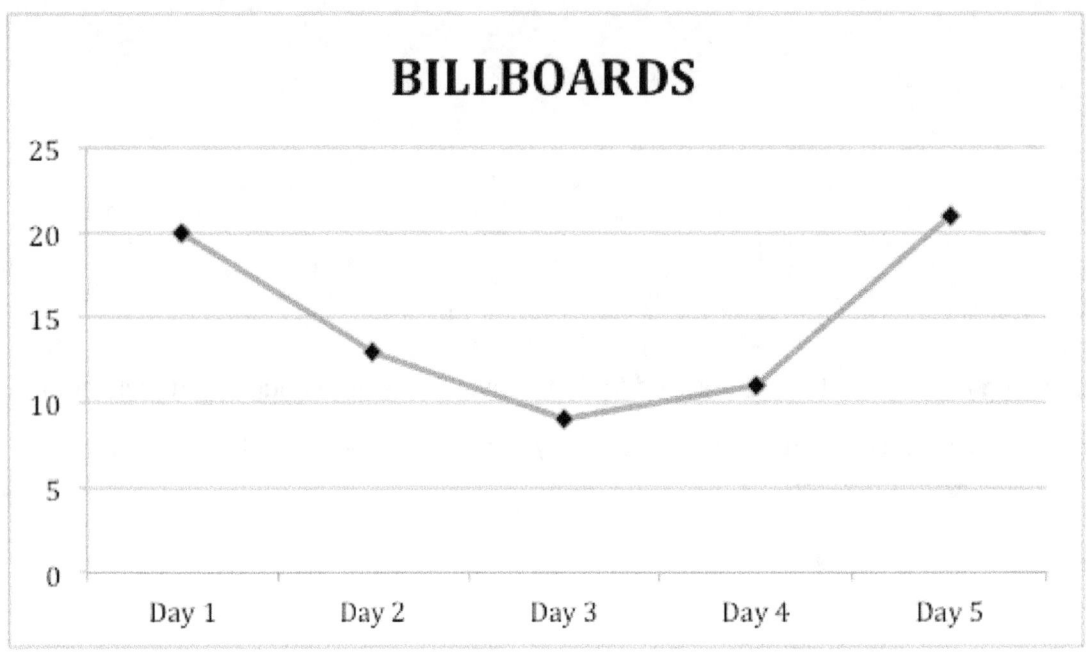

The above graph represents how many billboards a family counted per day on a five-day road trip. The following data can be gathered from the graph:

Day	Number of Billboards
1	20
2	between 10 and 15
3	just less than 10
4	just greater than 10
5	just greater than 20

Based on this data and the shape of the Line Graph, a few conclusions can be drawn:

1. The number of billboards counted on Day 1 and Day 5 are about double the number counted on the other three days.

2. The number of billboards counted dropped between Day 1 and Day 5.

3. Between Day 1 and Day 2, the number of billboards counted decreased.

4. Between Day 4 and Day 5, the number of billboards counted increased.

What could be one of the reasons for the decrease in the number of billboards counted between Day 1 and Day 2, a consistent lower number of billboards during Days 2, 3, and 4, and then an increase in the number of billboards between Days 4 and 5?

It is possible that the family traveled through a more rural area on Days 2, 3, and 4, and re-entered a more urban environment on Day 5.

Graphs – Practice Sheet

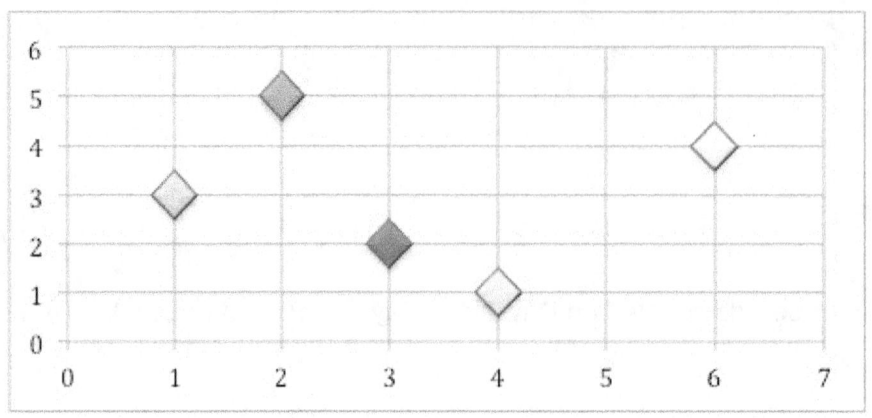

1. What type of graph is present above?

2. How many coordinates have been plotted?

3. Which color point has the greatest vertical value?

4. Which color point has 2 even coordinates?

5. Which color point is farthest left?

6. Which color point has the greatest horizontal coordinate?

Students Eye Color

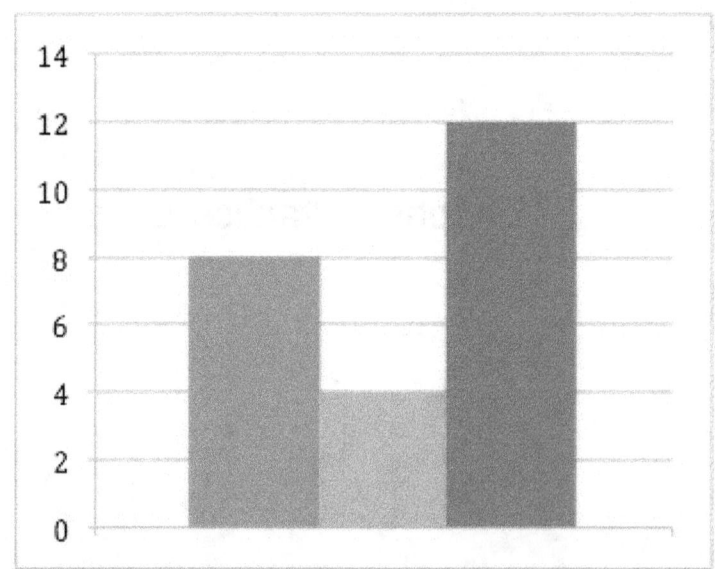

Based on the above graph, answer the following questions:

1. What type of graph is the graph above?

2. Which eye color do most students have?

3. How many more students have green eyes than brown?

4. How many more students have blue eyes than brown?

Types of Fruit Eaten at Breakfast for 7 days

Sarah	
Carlos	
Mason	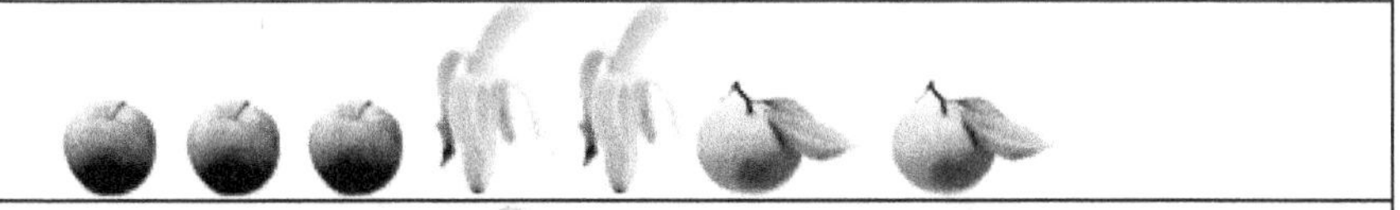
Karie	

1. What type of graph is presented above?

2. Which fruit did Sarah eat more of than the other students?

3. Which student ate the fewest bananas?

4. Which two fruits were consumed in equal total amounts?

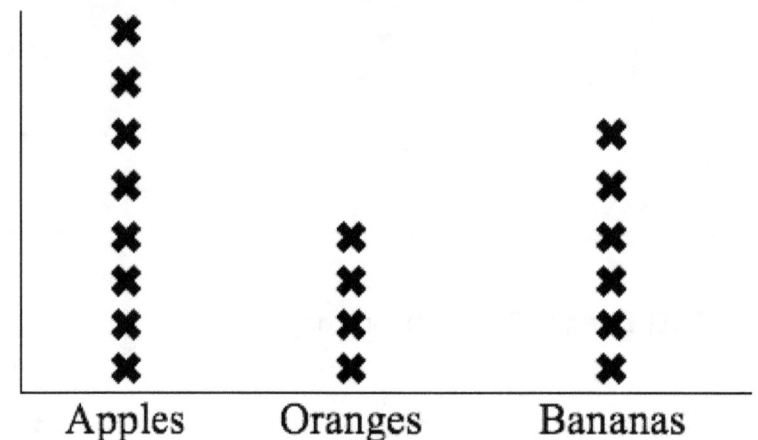

Students choice of Favorite fruit

1. What type of graph is shown above?

2. Which fruit is most popular?

3. Which fruit is least popular?

4. If each ✖ equals two students, how many students were polled?

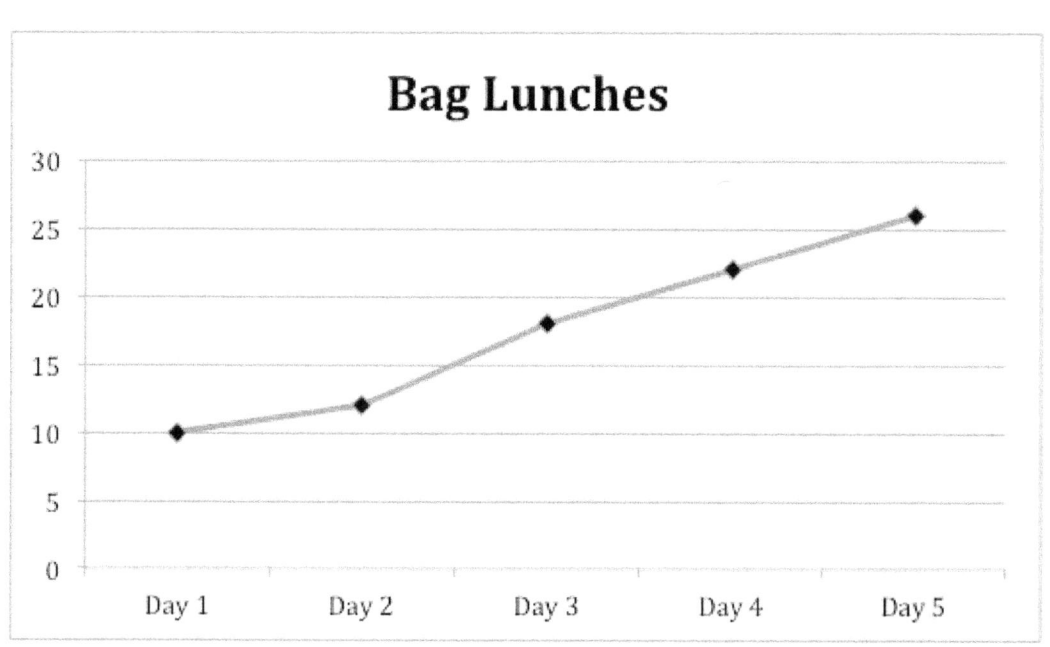

1. What type of graph is presented above?

2. Does the graph show an increase in bag lunches over 5 days or a decrease?

Graphs - Quiz

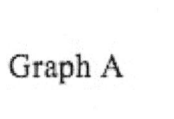
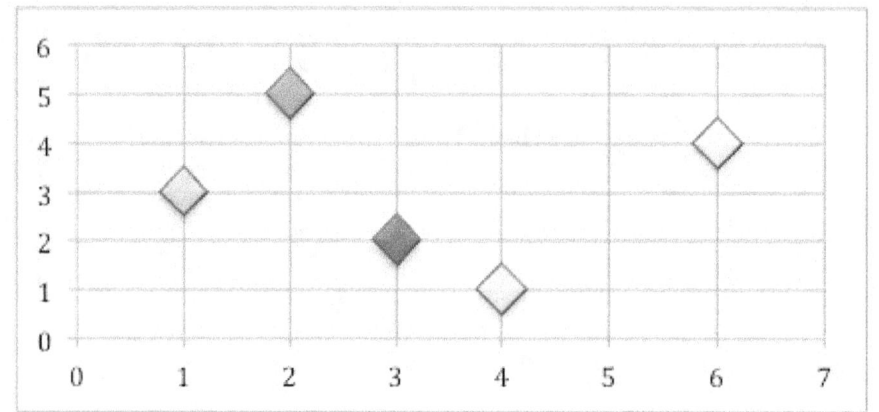
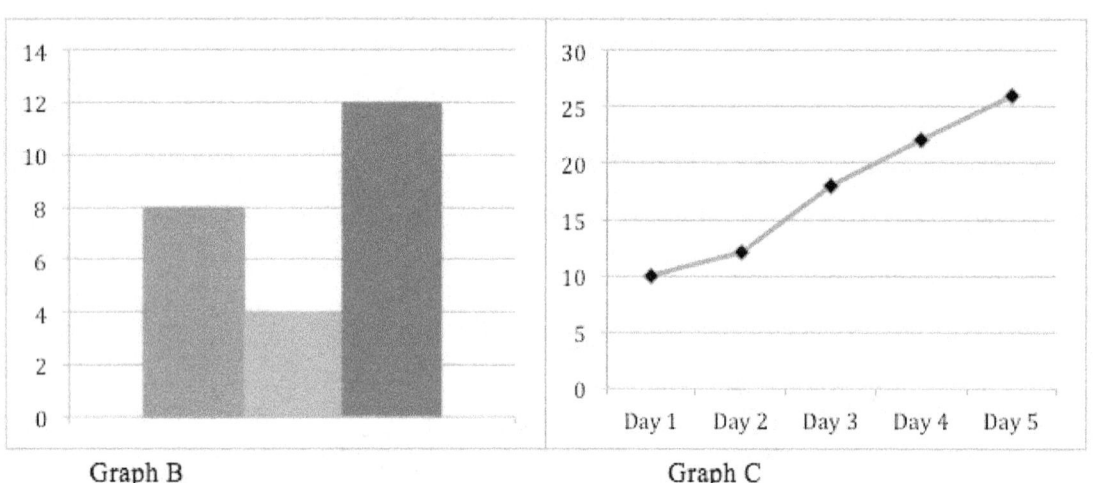

1. Which graph is a coordinate graph?

2. Which graph is a bar graph?

3. Which graph is a line graph?

4. Which graph shows a steady increase in data over time?

5. Which graph could show the square miles of water, flat land, and trees in an area to be developed?

6. How many points are plotted on the coordinate graph?

7. How many data groups are used on the bar graph?

8. Which two graphs could be converted one into the other easily?

9. Which two graphs previously discussed are missing?

10. True or False, graphs allow the student to visually asses a variety of data?

Answer Key

Number Sense – Practice Sheet

1. What is the place value for the number three in 3,526?   thousands

2. What is the place value for the number 7 in 743?   hundreds

3. What is the place value for the zero in 107?   Tens

4. Circle the whole numbers: 3, 4.3, 6, ½, 7.1, 9/10   Answer: 3 and 6

5. True or False: 4 3/4 is a whole number.   False

Number Names

1. Write the name for 205.   Two hundred five

2. Write the name for 1,706. One thousand, seven hundred six

3. Write the number for Three thousand, fifty-two.   3, 052

4. Write the name for 47th   Forty-seventh

5. Write the number for Eighty-Eighth 88th

6. Round 47 to the closest Tens Place: 50

Odds and Evens

1. Is 237 odd or even?   odd

2. Is 425 odd or even?   odd

3. Is 562 odd or even?   even

4. If you count by three's are all the numbers odd? No. every other number is   even.

5. If you count by four's are all the numbers even? yes

Comparing Numbers

1. Is 51 greater than or less than 15? Greater than

2. Which is the 3rd number after 10?   13

3. Which five numbers are between 12 and 18?   13, 14, 15, 16, and 17

4. How many zeros are in One Thousand?   3

5. How many tens (count by tens) are between Twenty and Fifty?   2 (30 and 40)

Number Sense - Quiz

1. Write the name for 2,358   Two thousand, three hundred fifty-eight

2. Write the name for 637   Six hundred thirty-seven

3. Round 84 to the nearest Tens Place: 80

4. Write the number for Seventy-Ninth   79th

5. Which numbers are in between 23 and 27? 24, 25, and 26

6. Which number is even: 14 or 17   14

7. Which number is greater: 23 or 32   32

8. Fill in the missing number: 2   4   ____   8   10   Answer: 6

9. If you come in 7th place in a race, how many people came in to the finish before you?   Answer: 6

10. Which number is less than 27: 26 or 28   Answer: 26

Addition and Subtraction - Practice Sheet

Fill in the missing numbers:

1. 1 + 3 = 4

    3 + ___ = 4   [1]

    4 – 1 = 3

    4 - ___ = 1   [3]

2. 2 + 5 = 7

    ___ + 2 = 7   [5]

    7 - ___ = 2   [5]

    7 - ___ = 5   [2]

Solve (remember to line up proper places or columns to get the correct answer)

1. 12 + 1 = [13]   2. 32 + 84 =   [116]   3. 810 + 21 =   [831]

4. 90 – 12 =   [78]   5. 98 – 15 =   [83]   6. 992 – 827 = [165]

7. 85 + 95 =   [180]   8. 410 + 127 = [537]   9. 321 + 620 = [941]

10. 87 – 18 =   [69]   11. 726 – 552 = [74]   12. 565 – 221 = [344]

13. 97 + 12 =   [109]   14. 123 + 456 = [579]   15. 722 + 32 =   [754]

Greater Than, Less Than, Equal To

1. 12 + 14 + 1 _____ 30 – 3   =

2. 125 – 18 _____ 100 + 2 + 3   >

3. 715 – 113 _____ 300 + 200 + 157   <

Addition and Subtraction - Quiz

Fill In the Blanks

1. 12 + _____ = 17, _____ + 12 = 17, 17 - _____ = 12, 17 – 12 = _____   [5]

2. 23 + _____ = 50, _____ + 23 = 50, 50 - _____ = 23, 50 – 23 = _____   [27]

Solve

1. 871 + 123 = [994]

2. 726 – 13 =   [713]

3. 35 + 27 =   [62]

4. 872 − 773 =   [99]

5. 73 + 271 =   [344]

6. 956 − 897 =   [59]

7. 25 + 29 =   [54]

8. 621 − 599 =   [22]

Fractions – Practice Sheet

1. How many pieces total are there in the above pie chart?   [8]

2. If you put the white piece and the darkest piece together what would be the   fraction? [2/8]

3. How many pieces equal 1/2 of the pie? [4]

4. What is the fraction for just one piece? [1/8]

5. If you picture the pie chart as a clock face, how many pieces equal 1/4 hour (15   min)? [2]

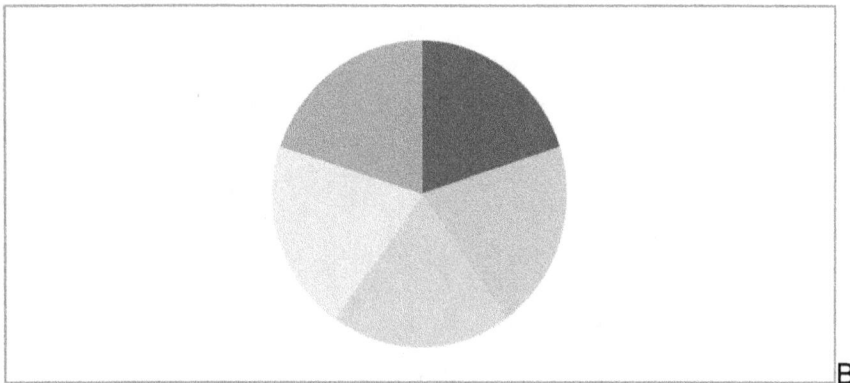
B

6. Is one piece from Chart A bigger or smaller than one piece from Chart B?   [smaller]

7. Which is bigger, 1/8 or 1/5?   [1/5]

Answer the following questions using Chart A and Chart B:

8. Which is bigger, 3/5 or 4/8?   [3/5]

9. If each chart were a real pie, which one would feed more guests? [A]

10. What is the denominator for Chart A? [8]   Chart B? [5]

Answer the following questions using the above set of shapes:

1. What is the total number of shapes?   [10]

2 How many squares are there? [5]

3. Write the number of squares and the total number of shapes as a fraction: [5/10]

4. How many shapes are not circles? [8]

5. What fraction of the total shapes are circles?   [2/10]

6. What fraction of the total shapes are triangles?   [3/10]

Compare Fractions

7. Put these fractions in order of smallest to largest: 2/3 2/7 2/5 [2/7,2/5,2/3]

8. Indicate whether the fractions are Greater Than, Less Than, or Equal: 7/7 ___ 3/3   [=]

9. Which Numerator is Greater: 5/16 or 9/10 [9/10]

10. Which Denominator is Greater: 1/8 or 4/7   [1/8]

Fractions - Quiz

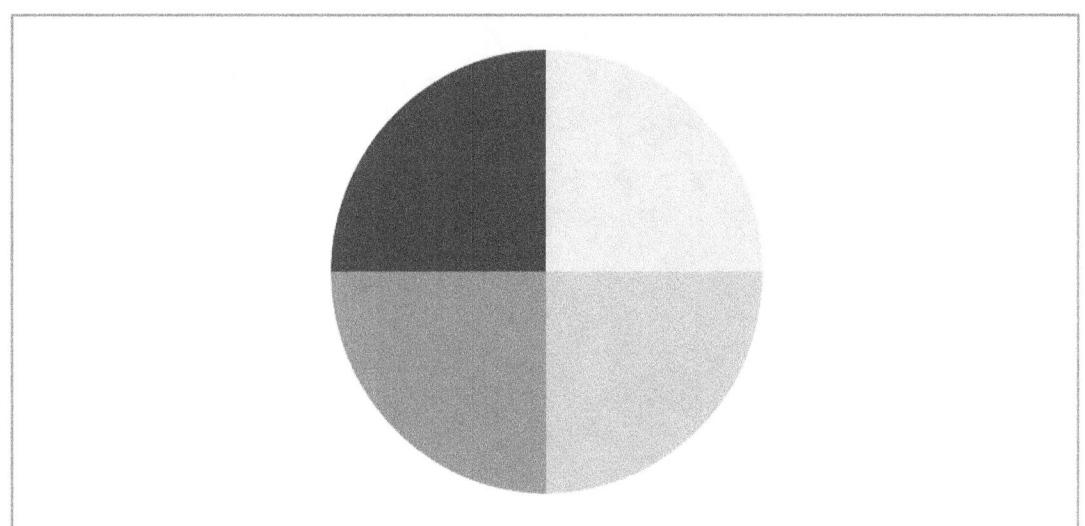

1. How many pieces are there in the above pie chart?   [4]

2. What fraction is the lightest color?   [1/4]

3. How many pieces make 1/2 of the pie chart?   [2]

4. What is the fraction for 3 of the four pieces? [3/4]

5. What denominator should be used for the above set of lightening bolts?   [6]

6. What is the fraction for the white lightening bolt? [1/6]

7. What fraction of the lightening bolts are gray?   [3/6]

8. What fraction of the lightening bolts are black?   [2/6]

Compare

9. Which is greater: 7/8 or 1/9   [7/8]

10. How many of the fraction 1/16 are needed to make a fraction equal to 1?   [16]

Operations – Practice Sheet

Place Values

1. What is the place value of the 8 in 185? Tens

2. What is the place value of the 9 in 19? Ones

3. What is the place value of the 2 in 207? Hundreds

4. What is the place value of the 4 in 4,123?   Thousands

Expanded Numbers

1. Write out the expanded number for 352:   300 + 50 + 2

2. Write out the expanded number for 85: 80 + 5

3. Write out the expanded number for 1,278:   1,000 + 200 + 70 + 8

4. Write the standard number for 20 + 9:   29

5. Write the standard number for 300 + 70 + 3:   373

6. Write the standard number for 5,000 + 200 + 20 + 1: 5,221

The clock above shows the hour hand on the 2 and the minute hand on the 12. The time represented is 2:00.

1. If it is during the day, is the time 2:00 AM or 2:00 PM?   2:00PM

2. If the minute hand is moved to the 6, is the time 2:15 or 2:30? 2:30

3. If the hour hand is moved to the 6, and it is after midnight but before noon, what time is it?

    6:00 AM

4. If the time is 8:15 PM, is it also 1/4 after 8:00 PM or 1/4 to 8:00 PM?

    1/4 after

Calendar

1. Which day of the week begins with the letter W? Wednesday

2. How many days are in the weekend?   Two

3. Which days are the weekend days?   Saturday and Sunday

4. Which day is next after Thursday?   Friday

Multiplication and Division

1. If there are 3 groups of 5 apples to be shared, how many people get an apple? 15

2. If there are 25 pencils and 5 students, how many pencils will each student get? 5

Operations – Quiz

1. What is the place value of the Zero in 1,027?   Hundreds

2. What is the place value of the 7 in 7,002?   Thousands

3. Write the standard number for 300 + 2 + 5:   325

4. Expand the number 7,329:   7000 + 300 + 20 + 9

5. Write an equivalent equation for 3 tens + 26 ones: 5 tens + 6 ones

6: If the time is 4:15 PM, what number is the hour hand pointing to?   4

7. If it is Tuesday, what is the name of the day to come next?   Wednesday

8. How many days per week do students go to school?   5

9. Since January is the first month of the year, what month is 4th?   April

10. Fill in the spaces: 5 x _____ = 50 AND 50 ÷ 5 = _____   10 for both

Money – Practice Sheet

Names and Amounts of Coins

1. Which coin has an amount of 25 cents? Quarter

2. Which coin has the amount of 1 cent?   Penny

3. What is the value of a dime?   Ten cents

4. What is the value of a nickel?   Five cents

5. How many quarters in One Dollar?   Four

6. How many dimes in One Dollar?   Ten

7. How many pennies in One Dollar?   One Hundred

8. How many nickels in one dime?   Two

9. How many pennies in one dime?   Ten

10. How many dimes in one quarter?   Two

Combinations and Amounts

1. How much money is 2 quarters and 3 pennies? $0.53 or 53¢

2. How much is 3 dimes, 2 nickels and 4 pennies? $0.44 or 44¢

3. How much is 3 quarters and 3 dimes? $1.05

4. How much is 2 Dollars and 4 dimes? $1.40

5. Which coins would be left if you started with 3 quarters and 3 dimes and spent 2   quarters and 1 dime?   1 quarter and 2 dimes

6. If you started with 6 dimes and 3 nickels and spent 2 dimes and 1 nickel, how   many quarters could you exchange the remaining dimes and nickels for? 50 cents would be left so 2 quarters.

7. If Tim started with $3.89 and bought a book for $1.23, what would he have left? $2.66

8. If Joy started with $3.25 and earned $2.00 for walking the dog and $1.00 for   setting the table, how much will she have to go shopping? $6.25

9. How much money will Sam have at the end of the week if he earns $8.75 and   spends $6.25? $2.50

10. If Josie starts with $9.85 and spends $7.38 on school supplies, how much would   she have left? $2.47

Money – Quiz

1. How many dimes equal the same amount as 6 quarters? 15

2. What is the name of the coin that is equal to 5 pennies?   Nickel

3. How many pennies in One Dollar? 100

4. How many nickels in 3 quarters?   15

5. What combination of quarters, dimes, and pennies is $0.98?

    3 quarters, 2 dimes, and 3 pennies

6. Write out the currency value for 3 quarters, 4 dimes, and 2 nickels: $1.25

7. If Danny earns $3.00 from his aunt, $2.50 from his Grandpa, and $1.25 from his   cousin, what would be the total amount of money Danny earned? $6.75

8. If Lisa has 6 quarters and 21 pennies, does she have enough money to buy a slice   of pizza for $1.75?   No, 6 quarters = $1.50 and 21 pennies = $0.21 and only   equals $1.71

9. If Lisa borrows $0.20 from her friend Luke, what will her change be after buying the pizza? $0.16

10. How much money will Brian have if he finds 2 quarters, 6 dimes, 4 nickels and 13 pennies?

$1.43

Shapes/Patterns - Practice Sheet

Define the following types of Patterns   [Answer: repeating, decreasing, and increasing]

Based on the above patterns, Answer the following questions:

1. What is the next shape in the first pattern?   square

2. What is the next shape in the second pattern?   circle

3. What is the next shape in the third pattern?   star

Fill in the Blanks

1.   5, 10, _____, 20, 25, 30 [15]   2.   2, 5, 8, ____, 14, 17, 20, 23 [11]

3.   2, 4, 6, 20, _____, 60, 200, 400, 600, 2,000, 4,000 [40]

4.   100,000, 10,000, _____, 100, 10, 1   [1,000]

Find the next number in each sequence

1.   1, 3, 5, 7, _____   [9]   2.   3, 6, 9, 12, 15, _____   [18]

3. 2, 4, 6, 8, _____   [10]   4.   35, 40, 45, 50, _____   [55]

Alternate numbers with the letters of the alphabet from A to Z

1. Count by 2's 2 A 4 B 6 C 8 D 10 E 12 F 14 G 16 H 18 I 20 J 22 K 24 L 26

2. Count by 3's 3 A 6 B 9 C 12 D 15 E 18 F 21 G 24 H 27 I 30 J 33 K 36 L 39

3. Count by 5's 5 A 10 B 15 C 20 D 25 E 30 F 35 G 40 H 45 I 50 J 55 K 60 L 65

4. Count by 10's   10 A 20 B 30 C 40 D 50 E 60 F 70 G 80 H 90 I 100 J

True or False

1. Counting by 2s represents an increasing pattern. TRUE

2. A countdown is considered to be a repeating pattern. FALSE

Shapes/Patterns Quiz

How many squares are repeated in the pattern? 2

How many triangles are repeated in the pattern? 1

How many circles repeat in this pattern?   3

How many diamond shapes repeat?   1

Complete the following patterns:

1.   1, 3, 4, 7, 11, _____ [17]

2.   2, 4, 6, 12, 14, 16, 22, 24 _____   [26]

3.   81, 72, 63, 54, _____   [45]

Fill in the blanks:

2. 5, 10, 15, _____, 25, _____, 35, 40   [20, 30]

3. 3, 7, 10, 17, 27, _____   [44]

True or False

1. If the following pattern continues, the next shape will be rectangle. FALSE

   Rectangle, square, triangle, rectangle, square, triangle, rectangle, _____

2. Counting by fives produces two patterns: an increasing pattern and a repeating odd and even pattern. TRUE

Length – Practice Sheet

Standard Length Units

1. Which is greater, 12 inches or 2 feet?   [2 feet]

2. Which is greater, 3 feet or 3 yards?   [3 yards]

3. Which is smaller, 12 inches or 1 foot?   (they are equal]

4. A Person's height is usually measured in _____.   [feet]

5. The height of a mountain is usually measured in _____.   [miles]

6. Which is greater 12 yards or 1 mile?   [1 mile]

7. Would a student measure his height using inches or miles?   [inches]

8. Which is a better unit of measure for determining the distance between two cities,   inches   or miles?   [miles]

9. How many feet are in 1 yard?   [3]

10. How many feet in 1 mile? [5,280]

Metric Length Units

1. Which is longer, 1 meter or 1 centimeter?   [1 meter]

2. which is better for measuring distances between cities, centimeters or kilometers?   [kilometers]

3. Which is better for measuring the length of a spoon, centimeters or meters?   [centimeters]

4. Which is better for measuring the length of a driveway, kilometers or meters?   [meters]

Comparing Standard and Metric Length Units

1. Which is greater, inches or meters?   [meters]

2 Which is smaller, feet or kilometers?   [feet]

3. Which Standard and Which Metric Units are best for measuring the distance   between San Diego and Los Angeles?   [miles and kilometers]

4. Approximately, how many centimeters are in 1 inch? [2 1/2]

5. Approximately, how many kilometers in 1 mile?   [2 1/2]

6. Approximately how many feet in 1 meter?   [2 1/2]

Length – Quiz

1. Which Standard Unit is better for measuring the distance between two windows of a house? [inches]

2. Which Standard Unit is best for measuring the length of fabric needed to re-cover several pieces of furniture? [yards]

3. Which Standard Unit is best for determining how far San Francisco is from New York City? [miles]

4. Which Standard Unit is generally used for measuring a person's height? [feet]

5. How many inches in 1 foot?

6. How many feet in 1 yard?

7. Which are similar in length, miles and centimeters or meters and feet?

    [meters and feet]

8. Which are used for measuring long distances, miles and kilometers or inches and centimeters?

    [miles and kilometers]

9. Approximately, how many kilometers in 1 mile? [2 1/2]

10. Approximately, how many centimeters are in 1 inch? [2 1/2]

Weight/Capacity – Practice Sheet

1. A fully loaded logging truck stops to be weighed. Will its weight be measured in tons or pounds? [tons]

2. Several students weigh themselves for a science experiment. Should their weight be recorded in pounds or ounces? [pounds]

3. A pizza parlor weighs the amount of shredded cheese it uses for each pizza. Do they use ounces of cheese or pounds of cheese per pizza? [ounces]

4. What is the order of the Standard Units for Weight from lightest to heaviest?

    [ounces, pounds, tons]

16 ounces (oz) = 1 pound (lb)

2,000 lbs = 1 Ton (T)

5. How many ounces are in 2 pounds? [32]

6. How many Tons will 6,000 pounds create? [3]

7. How many pounds will 48 ounces create? [3]

2 pounds are approximately 1 kilogram

32 ounces are approximately 1 kilogram

1 ounce equals approximately 30 grams

8. If a great big science book weighs 4 pounds, approximately how many kilograms does it weigh?   [2]

9. 2 kilograms equals approximately how many ounces?   [64]

10 A 2-ounce golf ball weighs approximately how many grams? [60]

11. Which units of weight (standard and metric) would be best for measuring how much a ball of yarn weighs? [ounces and grams]

2 cups = 1 pint

2 pints = 1 quart

4 quarts = 1 gallon

12. If 6 pints of cream are needed for a pie recipe, how many quarts should be used?   [3]

13. 20 gallons of punch are needed for a party. How many quarts can be used? [80]

14. How many cups will 15 pints of apple juice produce? [30]

15. How many pints equal one gallon? [8]

16. Which is the best unit of measure to determine the capacity of a swimming pool?   [gallons]

17. Which unit of capacity is best for measuring how much sugar is needed for a cake recipe? [cups]

18. Is it more likely to find quarts of juice at the store or cups?[quarts]

19. How many liters equal approximately 1 gallon?   [2]

20. Are gallons or liters of soda-pop sold in the average grocery store?   [liters]

Weight/Capacity – Quiz

1. Which of the above units represent the Standard Units for Weight? [ounces, pounds, tons]

2. How many Liters approximately equal 1 gallon? [2]

3. In an Ice Cream eating contest the winner ate a total of 8 pints. How many quarts did the winner eat? [4]

4. Twelve cups of pineapple are needed for a special recipe. How many 1-quart cans of pineapple are required? [3]

5. Approximately, how many liters of water will be needed to fill 8 1-gallon containers? [16]

6. A wheelbarrow carries a small pile of bricks to the backyard of a home to build a new fire pit. Should the bricks be weighed using pounds or ounces? [pounds]

7. Approximately how many pounds does a 5-kilogram package weigh? [10]

8. How many cups are needed for a recipe that calls for 6 pints of cream? [12]

9. A huge slap of stone that weighs 3 tons is going to be carved into a statue. The artist has to pay for the stone by the pound. How many pounds does he have to pay for?   [6,000 pounds]

10. If an item weighs 2 kilograms, approximately how many ounces does it weigh?   [64]

Probability – Practice Sheet

True or False

1. If an event is considered to be likely to occur, then it will definitely occur every time the test is conducted.   [false]

2. If an event is considered to be unlikely, then although it will not occur often, the possibility that the event could occur does exist.   [true]

3. When an event is certain to occur, there is still a small chance it will not occur on occasion.   [false]

4. When an event is considered to be impossible, it will never occur as long as the circumstances remain the same. [true]

More Likely, Less Likely, Equally Likely

1. A spinner used for playing board games is half red and half blue. What is the likelihood that the arrow will land in the blue area? [equally likely to land on red or blue]

2. During the winter months in Michigan, what is the likelihood that it will snow?   [more likely in winter than other months]

3. During the hot summer months in the desert, what is the likelihood of rain? [it is less likely to rain]

4. A bag full of jellybeans has 100 red jellybeans and 5 yellow jellybeans. Which color is more likely to be grabbed in a handful?   [red]

5. A deck of cards has exactly 4 kings and 4 queens. If only these 8 cards are used, what is the likelihood of drawing a king rather than a queen? {it is equally likely to draw a king or queen]

6. A teacher hands out pencils to her class to take a test. There are forty students and 50 pencils. If only two of the pencils are sharpened, what is the likelihood that a student will get a sharpened pencil? [it is less likely than getting a non-sharpened pencil]

7. A tennis champion has won 9 out of the last 10 matches against his opponent. Is it more likely or less likely that he will win the next match against this same opponent? [more likely]

8. If a second grade student reaches into a bag containing the last 10 chocolate bars and last 3 lollipops from Halloween, which sweet treat is more likely to be grabbed? [chocolate]

9. Mr. Smith has carved the Smith Family Thanksgiving turkey for the last 8 Thanksgiving dinners. It is more likely or less likely that he will carve the turkey again this year? [more likely]

10. Madison watches cartoon everyday after school for a half-hour before doing her homework. Is it more or less likely she will do her homework before watching cartoons tomorrow after school? [less likely]

Certain, Probable, Unlikely, Impossible

1. The sun rises in the east and sets in the west. What is the likelihood that it will rise in the west tomorrow morning? [impossible]

2. A robin lays three eggs in the spring. What is the likelihood that all three chicks will be baby robins? [certain]

3. A monarch caterpillar spins a cocoon and undergoes the change into a butterfly. What is the

likelihood that it will emerge as a Blue Morpho butterfly?   [impossible]

4.   A bag contains 25 black checkers and 1 red checker. What is the likelihood that a   black checker will be blindly chosen? [probable]

5.   A dozen chocolate chip cookies are made with semi-sweet chocolate chips. What   is the likelihood that there will be a cookie with white chocolate chips?   [impossible]

6.   A package of cherry popsicles is purchased at the store. What is the likelihood   that when opened the package will contain grape popsicles? [impossible]

Probability – Quiz

1. Which graph shows the possibility of two event outcomes?   [Chart 1]

2. Which graph shows either a certain outcome or an impossible outcome? [Chart 2]

3. In Chart 1, which color depicts the less likely outcome? {light gray]

4. What is the likelihood of an arrow hitting a white slice on Graph 2? [impossible]

5. What is the likelihood of a paintball striking the dark gray area on Chart 1? [probable]

6. If the light gray portion of Chart 1 was increased to cover half of the circle, would   the likelihood of it being struck by a paintball increase or decrease? [increase]

7. A box of crayons contains 20 purple and 20 orange colored crayons. If a student   reaches into the box without looking into it, what is the likelihood that an orange   crayon will be chosen opposed to a purple crayon? [the likelihood is equal of   choosing a purple or orange crayon]

8. A teacher passes out a spelling test. All the students passed the pre-test given the   day before. Is it more or less likely that all of the students will pass the test.?

[more likely]

9. A newspaper boy always throws the paper so it lands on the front porch of a specific house. When he delivers his paper tomorrow morning, what is the likelihood that the paper he throws lands on the front porch again? [probable]

10. As the earth circles the sun, seasons change in certain parts of the world. The cycle is winter, spring, summer, and fall. What is the possibility that summer will follow winter next year? {impossible}

Graphs – Practice Sheet

1. What type of graph is present above?   [coordinate graph]

2. How many coordinates have been plotted? [5]

3. Which color point has the greatest vertical value?   [red]

4. Which color point has 2 even coordinates?   [yellow]

5. Which color point is farthest left?   [violet]

6. Which color point has the greatest horizontal coordinate? [yellow]

Based on the above graph, answer the following questions:

1. What type of graph is the graph above? [bar graph]

2. Which eye color do most students have?   [blue]

3. How many more students have green eyes than brown?   [4]

4. How many more students have blue eyes than brown?   [8]

1. What type of graph is presented above?   [pictograph]

2. Which fruit did Sarah eat more of than the other students?   [bananas]

3. Which student ate the fewest bananas? [Karie]

4. Which two fruits were consumed in equal total amounts? [apples and oranges]

1. What type of graph is shown above? [line plot]

2. Which fruit is most popular? [apples]

3. Which fruit is least popular? [oranges]

4. If each ✖ equals two students, how many students were polled? [36]

1. What type of graph is presented above?   [line graph]

2. Does the graph show an increase in bag lunches over 5 days or a decrease?   [increase]

Graphs - Quiz

1. Which graph is a coordinate graph?   [Graph A]

2. Which graph is a bar graph?   [Graph B]

3. Which graph is a line graph?   {Graph C]

4. Which graph shows a steady increase in data over time? [Graph C]

5. Which graph could show the square miles of water, flat land, and trees in an area   to be developed?   [Graph B]

6. How many points are plotted on the coordinate graph?   [5]

7. How many data groups are used on the bar graph?   [3]

8. Which two graphs could be converted one into the other easily? [Graphs A and C]

9. Which two graphs previously discussed are missing? [pictograph and line plot]

10. True or False, graphs allow the student to visually asses a variety of data? [true]